DECORATIVE IRONWORK

some aspects of
design and technique

DECORATIVE IRONWORK

some aspects of design and technique

Published by
RURAL DEVELOPMENT COMMISSION
SALISBURY

Dear Reader:

We're very happy to bring the classic *Blacksamithing Series* from COSIRA (Council for Small Industries in Rural Areas) back into print. In the 35 years I have worked as a bookseller and publisher, certain iconic titles stick out. This series is one of those considered a classic because the photographs, text, and illustrations are so complete and highly focused on its subject.

The series:

- *Blacksmith's Craft* (978-1-4971-0046-6)
- *Wrought Ironwork* (978-1-4971-0064-0)
- *Decorative Ironwork* (978-1-4971-0063-3)

You hold the *Decorative Ironwork* volume in your hand. As you read, please be aware that we have not made any attempt to update the techniques or tools. Consider this treasure trove of knowledge a time capsule from the past.

May you be inspired to pick up the blacksmith's hammer and try your hand at this ancient skill.

Enjoy!

Alan Giagnocavo, Publisher
Fox Chapel Publishing

 This version published by Fox Chapel Publishing Company, Inc., 903 Square Street, Mount Joy, PA 17552.

ISBN 978-1-4971-0063-3

The Cataloging-in-Publication Data is on file with the Library of Congress

Printed in China
First printing

Important

A famous American humorist once described a certain type of film as 'films which begin in the middle for the benefit of people who come in in the middle'.

Unlike such films, this volume has been designed to be read from the beginning.

It is directed principally at craftsmen: men who start a job at the beginning and patiently carry on through their task with the determination to achieve results of the highest quality in the end.

We respectfully ask readers to

begin at the beginning.

CONTENTS

INTRODUCTION

This is the third manual on the subject of blacksmithing to be published by the Council for Small Industries in Rural Areas.

The first, *The Blacksmith's Craft,* dealing with basic and general smithing, was followed by *Wrought Ironwork,* a technical work in which the basic elements of the decorative side of the smith's craft were described.

In both these works great attention was paid to technical detail throughout the profusely illustrated texts. To a large degree they were complementary and were addressed to the novice as well as the more advanced smith.

This third volume has, however, been compiled more with the specialist ironworker in mind. Consequently, a number of operations included under the heading of standard practice have not been described in detail, as this would have proved unnecessarily tedious to the skilled smith, and would also have tended to divert the aim of the book.

The purpose of the authors was, broadly speaking, threefold.

First to introduce a number of methods of working iron not normally found in association with scroll work; methods in fact dissimilar to those commonly employed in this country since the early 18th century, but lending themselves more readily to designs akin in character to work executed prior to that period.

Secondly, to suggest to craftsmen by means of certain of these methods or techniques, that a return to a form of design in which flat metal surfaces, rather than narrow edges, are more prominently displayed, might tend to produce ironwork better suited to some styles of modern architecture.

In relation to the latter point, it was felt that the style of modern continental ironwork, in which similar technical methods have been exploited for some considerable time, would eventually be bound to influence patrons of the craft in this country. Until recently, designs copied or derived from the 18th century manner have been generally acceptable in Britain, but with the upsurge of continental travel the need for greater flexibility in ideas might arise.

This possibility materially affected the selection of techniques dealt with in this volume. They were chosen with the object of demonstrating to the craftsman and designer that quite a small number of 'moves' can yield a wider range of possibilities in design than might be expected. In this connection grilles Nos. 3 and 4 afforded an interesting example of identical methods producing, simply by the change in direction of a chisel cut, two designs similar in structural form but differing widely in ornamental effect. The inclusion of more examples of this kind was considered undesirable if the third aim of the book was to be achieved.

Thus, thirdly, it was hoped that the manner in which the technical material was presented would prompt those receptive of new ideas to enter the experimental field through the medium of design, and create for themselves.

The exploitation of a technique for its own sake alone is liable to give rise to departures from established practice merely to achieve novel or ingenious effects: such results are without merit and it is for the avoidance of this fatal error that the design factor is stressed here and at intervals throughout the text.

Though working drawings of each grille have been produced and are available to rural smiths, the grilles should not be regarded simply as catalogue designs, but rather as examples to be studied one in conjunction with another.

Before any attempt is made to use an individual text as a working recipe, the volume should be read, as initially intended, from cover to cover. For good reasons certain details were dealt with at the ends of the chapters, while other material points mentioned in the text applying to one example may throw additional light on processes applicable to one or more of the other designs. Thus it was hoped that the reader would, by following the text consecutively, develop a mental picture of the subject as a whole, before becoming preoccupied with specific details.

The examples have been called 'grilles' for convenience and not because the repeating designs employed are suitable for the making of grilles only, or to suggest that the decorative features must necessarily fill any given framework. Several of the devices might well be used sparingly in large gates, for example, in the form of borders or panels, so long as the overall character of the work is maintained; in fact the techniques dealt with in this book can be applied over the whole field of decorative ironwork.

Design 1

A feature of this design is the exploitation of contrasting bar sections. The use of rectangular section for one series of bars, and a round section for the other series, not only appeals to the eye, but also simplifies the making of the apertures in the flat bars through which the round bars pass.

The method employed eliminates the hot punching, drifting and dressing associated with intersecting bar work, and consequently saves time.

The lack of ability to fire weld is no handicap in the construction of this particular grille as the simple leaf forms are not made separately and attached, but are developed from the parent bar by methods which have, particularly since the 18th century, largely fallen into disuse in this country.

It will be appreciated that this design may be carried out in a wide range of bar sizes. Because the use of the chisel gives rise to a limited amount of drawing of the metal, and since the length of slit required to form an eye of given size must be determined, definite bar sizes are quoted and dimensions given. These figures are used solely for the purpose of rendering technical principles clear, and apply to $1\frac{1}{2}'' \times \frac{1}{4}''$ and $\frac{1}{2}''$ round bar.

The frame size of this particular example is $3' \, 9'' \times 2' \, 10\frac{1}{2}''$ and is made from $1'' \times \frac{1}{2}''$ bar.

DESIGN 1

Fig 1

The flat bar is marked out with chisel and centre punch. The drawing, Fig. 16 on page 16, gives the measurements for setting out when using $1\frac{1}{2}'' \times \frac{1}{4}''$ bar.

This marking, besides being accurate must be sufficiently indented to be seen clearly when the metal has been heated for slitting and cutting operation.

Fig 2

A slit is cut in the hot metal with a chisel $\frac{15}{16}''$ wide, this being the slit necessary for the formation of an eye through which $\frac{1}{2}''$ round bar will pass.

A cutting plate, preferably of copper, is used to protect the chisel edge from the hardened anvil face.

Fig 3

Using a $\frac{1}{2}''$ top fuller in conjunction with 1″ bottom swage, one half of the eye is formed.

A bottom swage of ample depth must be used if pinching-in the bottom of the tool and consequent malformation of the shape is to be avoided.

The bar is turned over and the other half eye is formed.

Assuming that no unnecessary violence has been used, it will be found in practice that no appreciable alteration occurs in the distance between centres, which in the present case is $7\frac{1}{2}''$, as shown in the diagram on page 16.

It is important to use a hand length of $\frac{1}{2}''$ round bar to check that eyes have been opened sufficiently to enable the longer round bars to pass through a series of eyes easily when the work is assembled.

Note: The setting of the eyes alternates. A scrutiny of the illustration of the finished grille will make this point clear.

Fig 4

A curved chisel is used to release the tip of the spur, and the straight cut is continued with a sharp hot-set.

It is here, in this operation, that a slight lengthening can occur. To minimise this effect, the hot-set must be thin and a good edge must be maintained.

From tip to butt the spur must be released progressively with well-controlled blows of a moderate weight.

In order that the natural chamfer made by the hot-set may appear uniformly on the face side of the work, all cutting must be done from one side only.

From time to time work must be checked to ensure that the measurements between the centres of the eyes is uniform.

Fig. 1

Fig. 2

Fig. 3

Fig. 4

DESIGN 1

Fig 5 The reason for the next step may not be obvious at this stage. It will suffice to state here that on it depends the correct formation and setting of the spurs. By carefully following the instructions for the succeeding operations the reason for this first step will become apparent.

The spur is given a short 90° twist bringing its outside edge uppermost in relation to the face side of the bar. In order to facilitate this operation the spur is first pulled out at a convenient angle.

Fig 6 The 90° short twist is made and this brings the curve of the spur tip into a position which assists when further curving the feature as forging proceeds.

The bar is gripped in the vice with the eye inserted a short distance within the jaws. This ensures stability and prevents distortion.

Fig 7 Using a curved faced hammer the twist is dressed to blend with an even flow into the main bar. If necessary the heat should be localised by controlled quenching to avoid distortion occurring either in the eye or the centre stem.

Fig 8 The twisted root of the spur is dressed on two sides roughly at right angles to one another in order to produce a uniform section. As it is clearly impossible to continue on the anvil bick, a stake with an acute angled flat is brought into use.

Fig 9 The spur is now curved over the bick, care being taken to avoid continuing this operation too far, as further curving takes place naturally during the stages to follow.

It is also convenient at this point to refine the shape of the spur tip a little, bearing in mind that a few light blows will suffice, as any tendency to depart from the leaf form is out of place in this design.

Fig 10 With the spur resting flat on the anvil both edges are hammer chamfered. A curved faced hammer is used for treating the inner

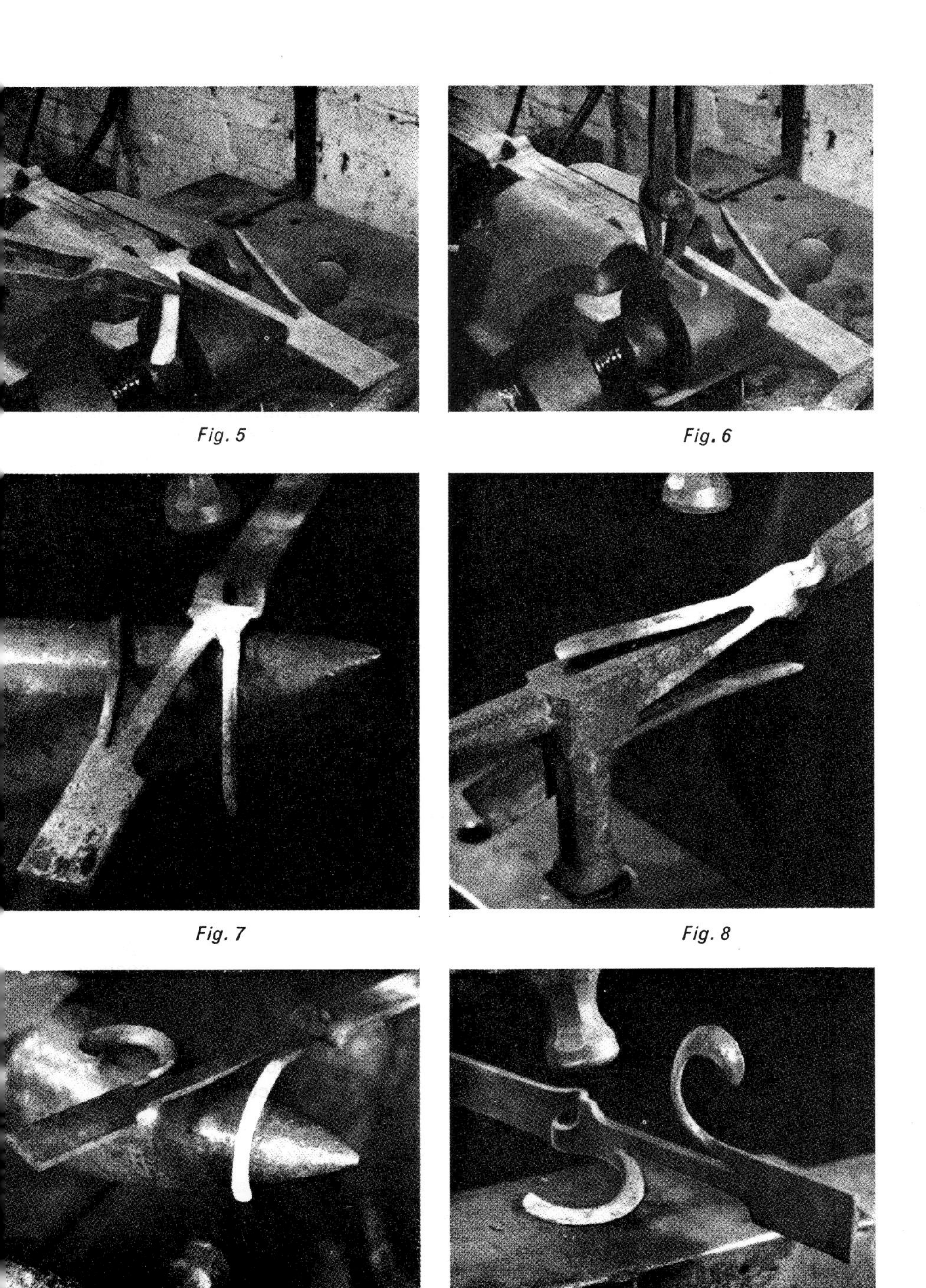

Fig. 5

Fig. 6

Fig. 7

Fig. 8

Fig. 9

Fig. 10

DESIGN 1

edge and a flat faced hammer for the outer, and the transformation from the spur into the leaf form develops.

Fig 11 The leaf form is still lying in a place at right angles to the face of the bar.

At this stage, however, the axis of the leaf is canted by resting the inside edge on the bick. The outside edge is therefore raised a little and is hammered down on to the bick with light coaxing blows, great care being exercised to avoid solid blows which would thin the metal and tend to distort the emerging shape.

This action is used progressively from tip to butt, the work being moved round the bick as shaping proceeds.

During this operation the leaf begins to revert to its original plane, moving back through the 90° setting described in Fig 5.

Fig 12 Additional work is still required to bring the leaf right back into a plane true with the parent bar.

This is done by gripping the top edge of the leaf with the round-nosed pliers at a point where leverage applied in a downward direction will bring the leaf into its correct setting.

This motion achieves two objects. First the leaf is set in its correct plane, and, secondly, the top edge is correctly canted and blended with a smooth sweep into the parent bar to complete the flow of the whole form.

Fig 13 After a pair of leaves has been forged they are given the customary inspection and any small adjustment is made. Nothing more than a light tap here and there with the hand hammer will usually be necessary.

Figs 14 and 15 Assuming that the frame has been made to the required dimensions, and all the decorative bars have been forged, work preparatory to assembly begins.

Using the full size working drawing for guidance, the forged bars are placed accurately on the frame and round bars are temporarily passed through the eyes to check alignment. The cutting-off points are marked at the correct angles on the flat bars where junction is made with the frame.

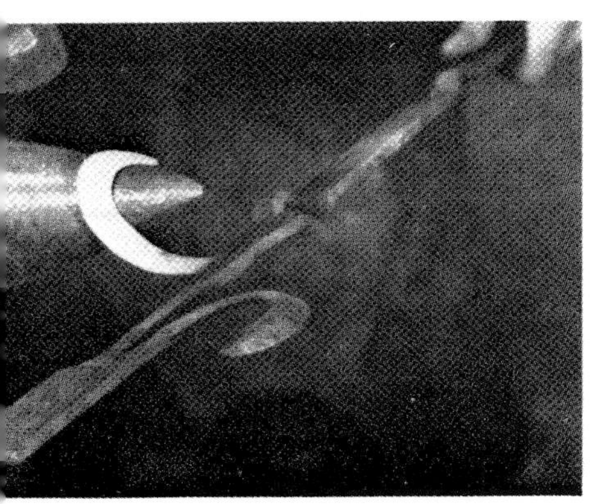

Fig. 11

Fig. 12

Fig. 13

Fig. 14

Fig. 15

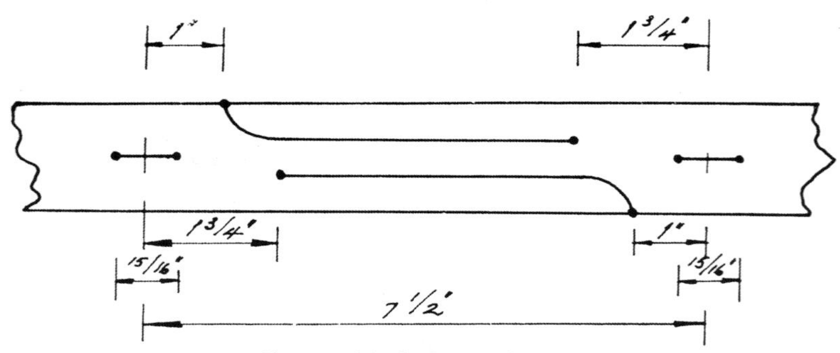

Fig. 16 Marking out diagram

After cutting to length the ends are shaped and hammer chamfered.

The positions of rivet holes are now marked and clearance holes drilled.

The bars are reassembled on the frame over the working drawing, and the rivet holes are scribered through on to the frame. These holes are now drilled and slightly countersunk at the back.

The bars are fixed into position temporarily with bolts and nuts which are replaced eventually by rivets.

For ease of working, the round bars which have forged bosses drilled for riveting at either end, are each made in two sections and oxy-acetylene welded together when in place. The butt joints must be arranged to fall in convenient positions between the decorative bars.

Where the bars of the grille abut the frame, a number of half squares occur naturally. They do not afford sufficient space to accommodate the leaf motif in the form used in the complete squares. Consequently a leaf of a different shape was designed to fill these triangular spaces. These leaf forms were produced from spurs released from the bar in the same manner as the main leaf motif, and were worked in a similar way, but were shaped to suit the proportions of the space to be filled.

To ensure the full effect of this design reasonably precise adherence to the full size working drawing must be maintained: nevertheless work should be freely forged and the direct character of the hammer work should not be spoiled by unnecessary fettling.

When assembling it may be found helpful to use a round file for easing the sides of the eyes through which the straight ½" round bars pass, should adjustment of alignment be necessary.

As one of the fixing holes in each of the flat bar-ends is also used for the fixing point of a round bar, the ends of the latter are set and cranked so that their rivet holes, when drilled, will be central in the bosses and coincide with the appropriate holes.

If it is desired to use this design for a gate, the complete grille could be housed within another frame made in the usual manner for gates, carrying the appropriate fittings. This would become a shadow surround, the grille being secured by countersunk screws.

Design 2

The design of this example deviates from the orthodox in the way that the decorative features are evolved from the structural bars of the grille and are not additions in the form of branch welded or collared scrollwork.

Such scrollwork and other applied ornamental devices have been used for generations to form panels and borders within vertical and horizontal straight bars. In earlier times, however, the splitting of sections of metal was employed to a greater extent than has been practised since the 18th century, to relieve, or partially release from parent bars, portions which were fashioned into decorative features or points of attachment.

In the present example this technique has been employed in the fashioning of the intermediate bars to achieve a decorative effect without resorting to the use of applied embellishments.

Splitting is also exploited in the branching leaves of the freely forged cresting.

The wavy bars, built up from forged shapes fire welded together, also embody their own decorative elements within the bar without recourse to the addition of supplementary motifs.

To achieve the strongest visual effect the frame is constructed with the broad faces of the bars outwards and the internal bars are secured by means of rivets, which, in themselves, contribute something of value to the general effect of the design.

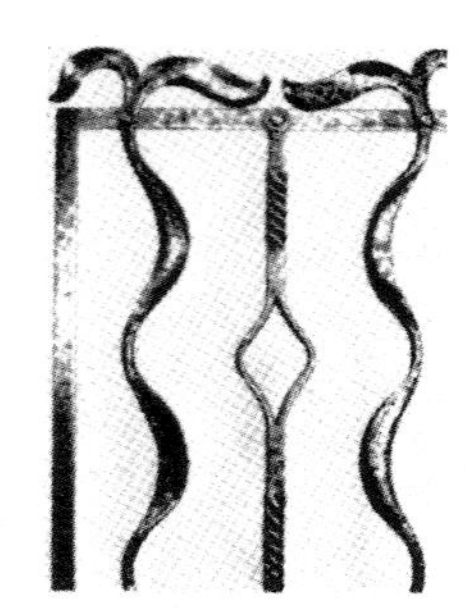

DESIGN 2

The split-forms in the alternate bars are not equally spaced. The intervals between them are carefully proportioned to give life to the design and avoid a feeling of heaviness.

To give a play of light and shade the concave curved sections of the wavy bars alternate both horizontally and vertically.

Where dimensions occur in the text they refer specifically to the example illustrated, the frame of which is made of $\frac{3}{4}'' \times \frac{1}{2}''$ bar and measures 3′ 3″ × 2′ 7″. These dimensions would, however, be adequate for larger areas, though enlargement would tend to lighten the general effect. This would not necessarily be a disadvantage.

A scale drawing should always be made to confirm the suitability of the proportions of the proposed design for any given site.

Fig 17 Assuming that a start is made on the split features, a bar, in this instance $\frac{5}{8}''$ square and long enough for making a complete upright to the dimensions given in the working drawing, is cut off.

After marking out the bar is cut with a hot-set almost right through to its other side to the required length.

The work is turned over and the split is completed. It is advisable to radius very slightly the corners of the hot-set blade.

Fig 18 The split is opened a little with the hot-set to allow an elongated drift to be inserted. A bolster is used to avoid damaging the edge of the hot-set on the anvil face.

Fig 19 The split is now ready to accept an elongated drift.

Fig 20 The drift and deep bolster with which the next stage of the work is done. See drawings, Figs 46 and 47, page 29.

Fig 21 The work is reheated and the drift is driven into the split but not to its full extent.

Fig 22 Both faces of the bar are trued with the hand hammer. After dressing one face the work is turned over and the other face is treated.

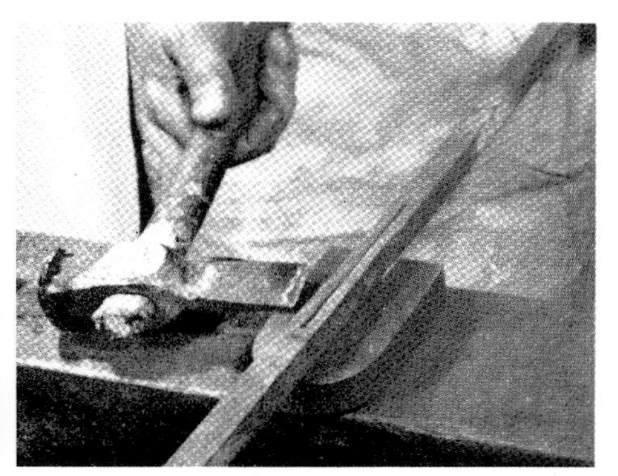

Fig. 17

Fig. 18

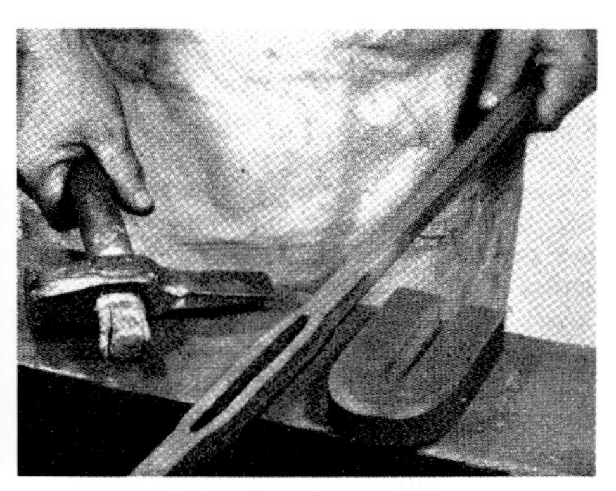

Fig. 19

Fig. 20

Fig. 21

Fig. 22

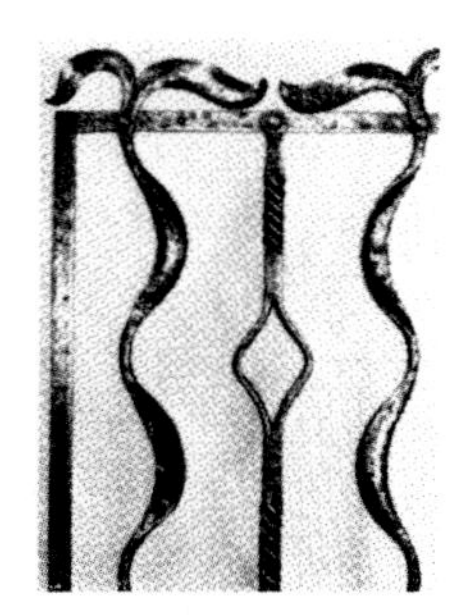

DESIGN 2

Fig 23 The drift is now driven home to its full extent and the truing process is completed.

Fig 24 To make the diamond shape the split is opened in the middle as a lead-in to the next operation.

Fig 25 Another heat is taken and the bar is hammered up from one end to widen out the naturally formed diamond shape to the required extent. The effect of the blows should be controlled by turning the work over occasionally. If this point is neglected serious distortion may develop and the freshness of the work may be lost in course of adjustment.

Fig 26 The shape of the diamond feature is now finished on the anvil bick to conform with the contours of the full size working drawing. A delicate touch with the hammer is called for here.

Fig 27 To make the circular opening, the initial splitting and dressing is carried out as before; but the first stage in the actual opening of the feature differs.

In this case both ends and the centre of the split are widened by the insertion of the hot-set. This operation predisposes the split to assume its final ring-out form.

Fig 28 After reheating, the bar is hammered-up from one end and the process started in the previous operation is further developed and the split nears its final form.

Fig. 23

Fig. 24

Fig. 25

Fig. 26

Fig. 28

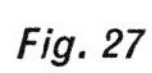

Fig. 27

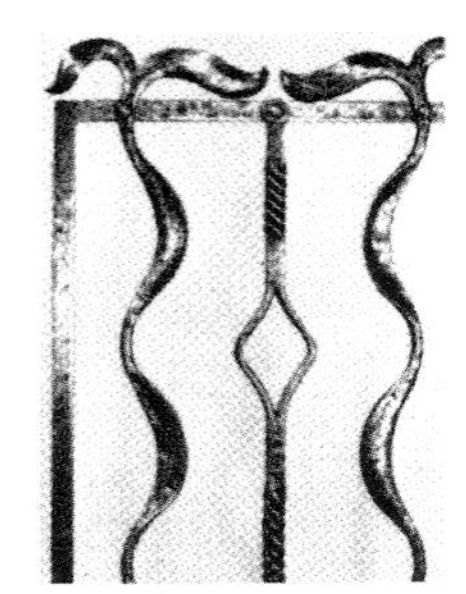

DESIGN 2

Fig 29 The ring feature is now fully opened across the anvil bick by hammering on each end in turn after heating and reheating.

The shape is finally adjusted to conform with the working drawing.

Fig 30 The twisted sections forming part of the embellishment of these bars are now prepared.

The length of the section to be twisted is marked with centre dots.

Top and bottom $\frac{5}{8}''$ rope tools are used to indent and round the bar in such a manner as to convert the square section into a quatrefoil section; thus an assembly of four $\frac{5}{16}''$ round rods is simulated.

Fig 31 An increase in length occurs during the operation described in Fig 30, owing to the drawing effect of what is virtually a swaging action.

The prepared section is twisted at a moderate, even, red heat (see page 46, Fig K of *Wrought Ironwork**) and this process restores it to approximately the original length.

Fig 32 Care is taken throughout to ensure that the twist ends precisely at its junction with the plain square section of the bar.

Fig 33 The successful forging of the wavy bars calls for a dishing block, a special tool in the form of a simple die (see drawing, Fig 48, on page 30). The shape of the die is obtained from the working drawing, but to furnish a better idea of what is required a full-scale plasticine model of one of the dished and curved shapes is useful. The impression in the tool is sunk by means of suitable fullers with the mild steel block at a bright heat. A forging is made from $\frac{3}{4}'' \times \frac{3}{8}''$ bar, the ends of which have been reduced and rounded leaving a centre swelling. This is the opening move in the production of one of the component shapes and it is here that the smith's judgement, as well as measurements, must play an essential part. The forging is left at this stage and retained

* CoSIRA Publication No. 55.

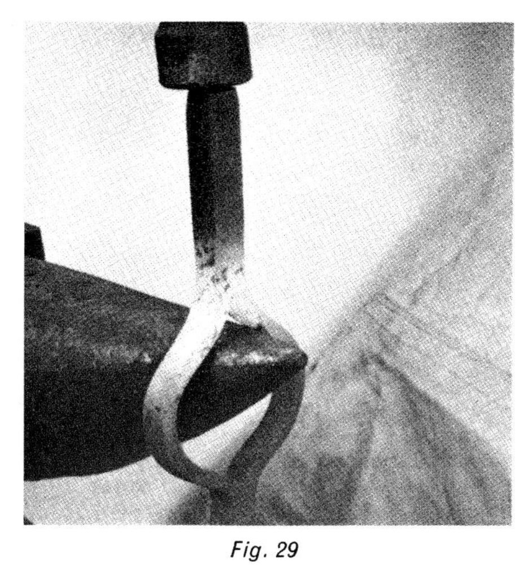

Fig. 29

Fig. 30

Fig. 31

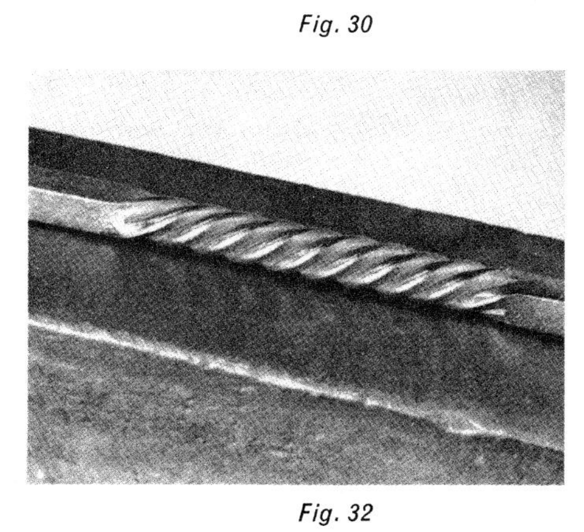

Fig. 32

Fig. 33

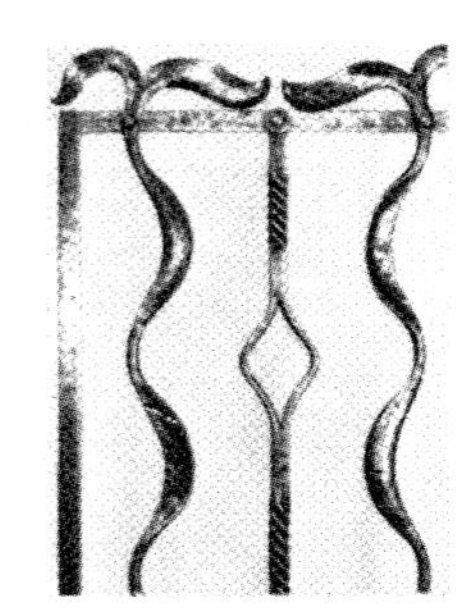

DESIGN 2

as a guide. Its value as a final pattern will be checked by the operations which follow in Figs 34 to 39, and in any of these succeeding stages the need for adjustment may be revealed. A handling length of $\frac{3}{4}'' \times \frac{3}{8}''$ bar is taken and, working to this trial pattern, the first stage is forged: it is, however, not severed from the stock bar.

Fig 34 Using anvil horns and scroll wrench the curve is now put in to conform with the impression in the tool.

Figs 35 and 36 The rectangular section of the blank is transformed into a shallow half-rounded section, first by shaping the outer edge with the normal hand hammer, followed by similar shaping on the inner edge, but this time with the curved faced hammer.

Fig 37 With the flat side uppermost the work is now dished in the tool. Suitable fullers are applied progressively round the curve, the smith's mate striking with moderate blows to avoid undesirable thinning occurring as the form is worked into the cavity.

Fig 38 The plane of the forging is trued. The outer edge will make contact with the anvil face leaving the inner edge raised naturally.

Fig 39 A chalk tracing of the appropriate portion of the working drawing is prepared on a steel sheet (see *Wrought Ironwork,* pages 12 and 13, Figs 12 to 15*) and the work is now set to follow the lines of this drawing.

The freshly forged piece is cut from the bar and work continues on the making of the requisite number for welding up into a decorative bar.

* CoSIRA Publication No. 55.

Fig. 34

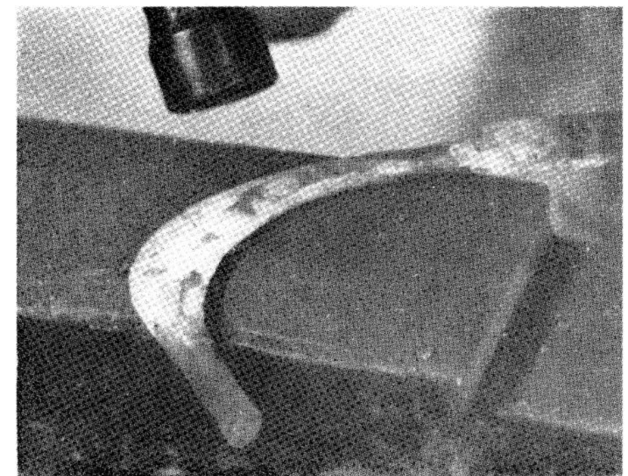
Fig. 35

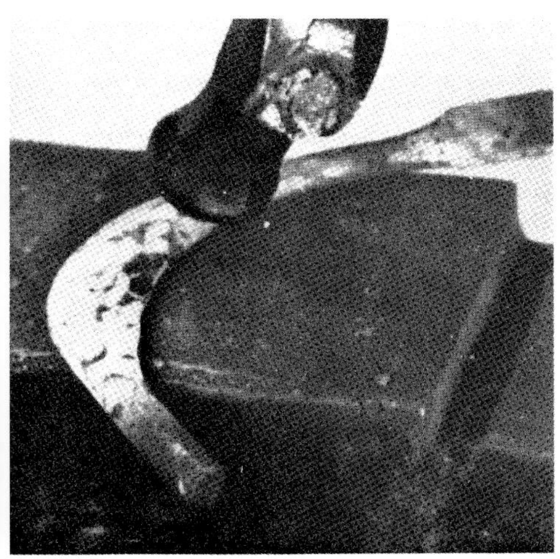
Fig. 36

Fig. 37

Fig. 38

Fig. 39
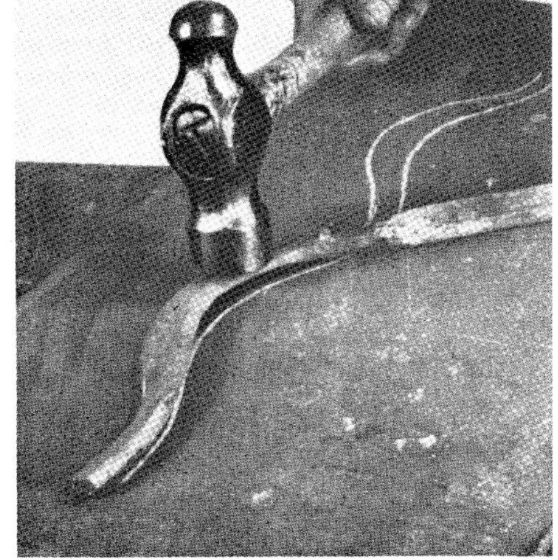

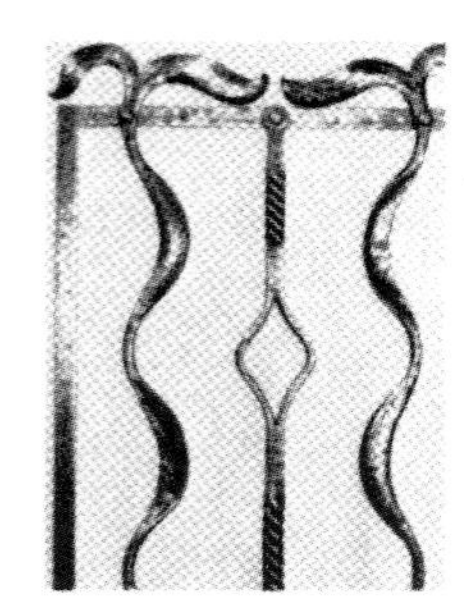

DESIGN 2

Fig 40 The forged shapes are fire-welded together with convex and concave sides alternating. In the example shown the welding process was arrested in order to show the position of the scarfs. Normally this weld is completed in one heat and dressed to the correct round section.

As each weld is finished the work is checked against the chalk tracing and set as required.

Fig 41 The pairs of branching leaves forming the cresting are forged, in the present instance, from $1\frac{1}{4}'' \times \frac{5}{16}''$ bar.

A heating length of the bar is split to the required depth. The cut is made off centre to produce lobes of unequal size, to conform with the design.

Fig 42 Immediately in wake of the split the bar is reduced by the use of a cheese fuller with the work positioned on the anvil bick.

Fig 43 The leaf forms are fashioned by short pointing the lobes and curving the points on the tip of the anvil bick.

Fig 44 The leaves are hammer-chamfered on all face side edges. This process, besides providing the desired surface texture, also spreads and shapes the leaves to conform with the design.

Too great an insistence on precision should be avoided here, as the leaves are freely forged and undue dressing or smoothing of the surfaces will destroy their character. A rivet hole is punched in readiness for fixing the bar on to the frame. The branching leaves are severed from the handling length leaving a shank long enough to fire-weld on to the top end of a wavy bar.

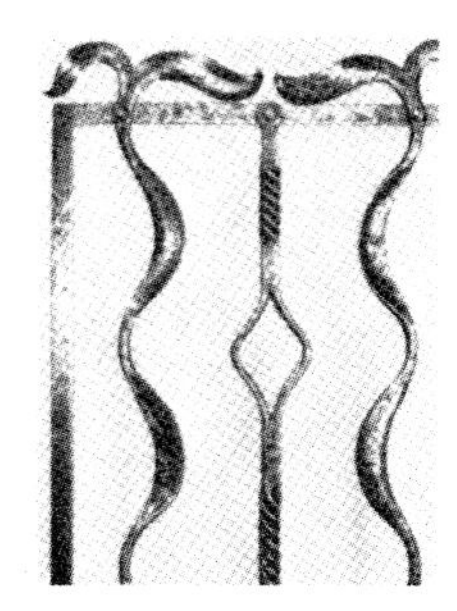

DESIGN 2

Fig 45

All decorative bars are riveted to the frame. The top rivet holes of the wavy bars have been mentioned, leaving the bottom holes and both rivet holes in the other type of bar to be dealt with.

In the case of the lower holes in the wavy bars, a swelled eye was produced at the end of the bar by slot punching and drifting to the appropriate size. The holes in the other bars are made differently. First the ends of the bars are rounded laterally. After the bar end is heated, a round punch with a $\frac{7}{8}''$ diameter flat face is positioned on the end of it. The punch is dealt a heavy blow by the smith's mate using a sledge hammer, and produces a flat round boss stepped-in from the face side of the bar, with an area and section of sufficient strength to permit the drilling of a rivet hole. If it is desired, distinction may be added to the riveting boss by a slight and graceful reduction of the square bar immediately adjacent to the boss. This refinement should, of course, be made over the anvil bick with the curved faced hammer before the boss is formed. Both operations described cause the bar to lengthen a little, consequently a reduction must be made prior to cutting away surplus bar at each end.

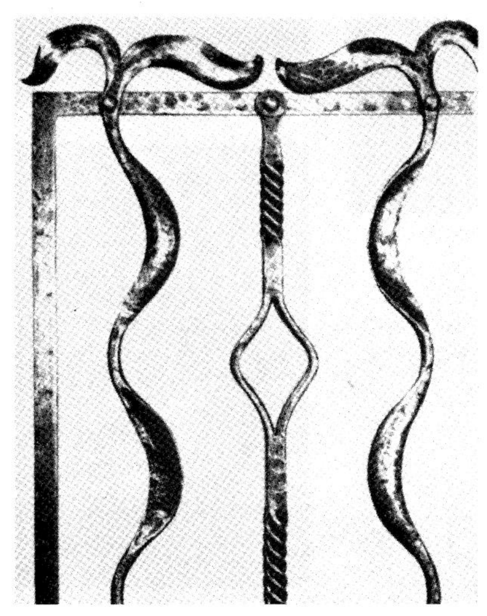

Fig. 45

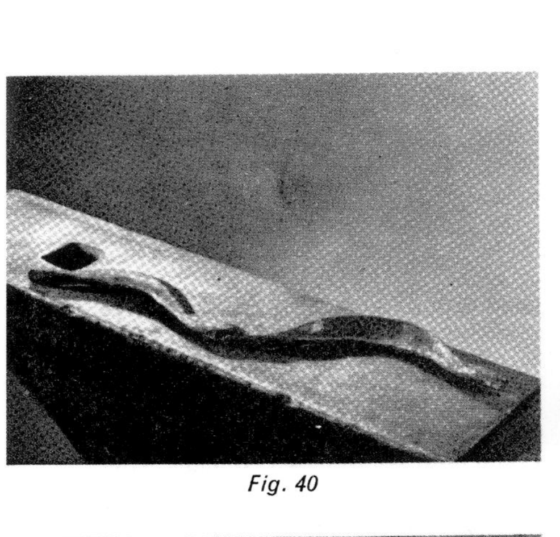

Fig. 40

Fig. 41

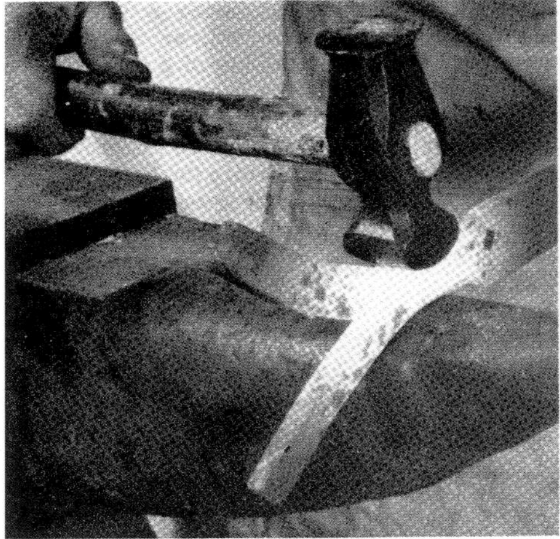

Fig. 42

Fig. 43

Fig. 44

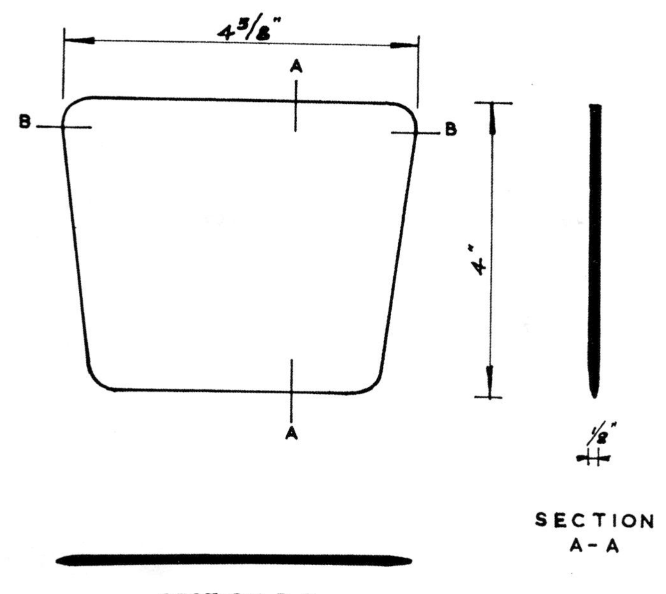

Fig. 46 Elongated drift

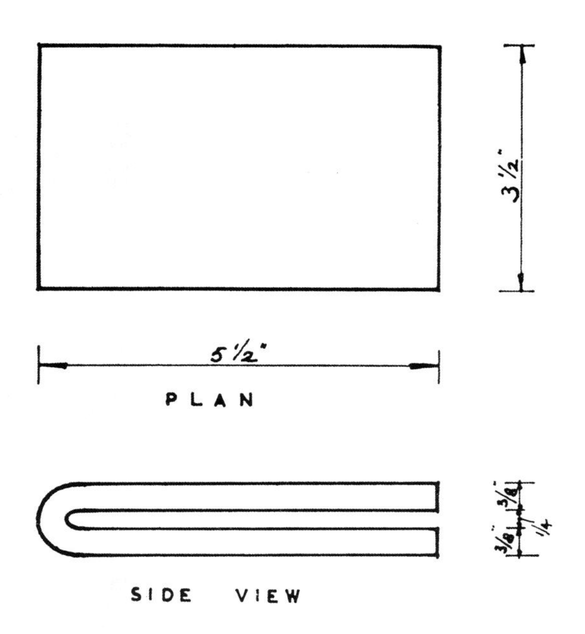

Fig. 47 Deep bolster

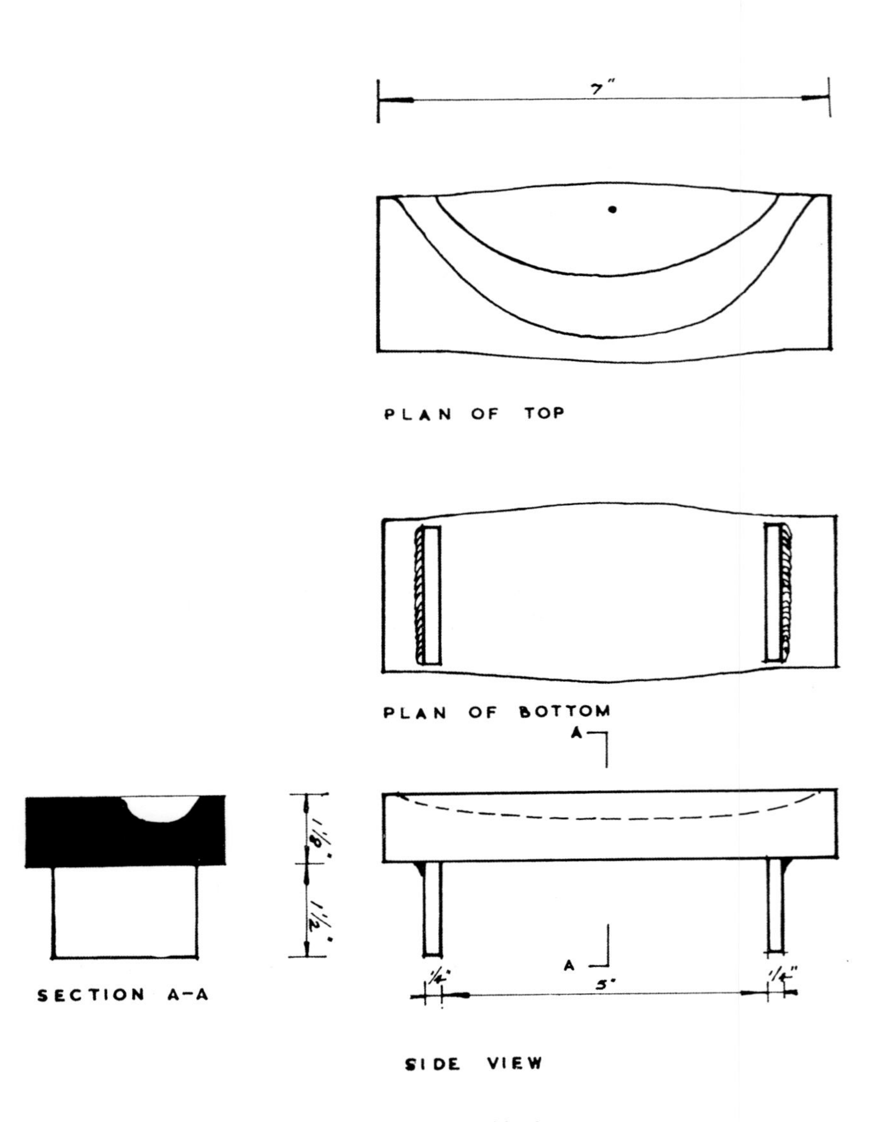

Fig. 48 Dishing block

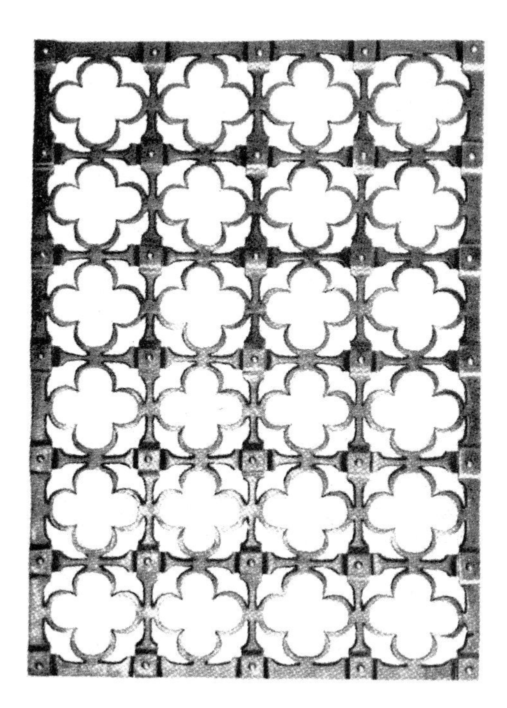

Design 3

This is an example of the time-honoured quatrefoil theme but one in which the decorative elements are formed almost entirely by means of a splitting technique.

Curved spurs, relieved from the bars by chiselling, form sections of each quatrefoil, leaving the parent bars with modified contours which play a part in the decorative whole at the points of intersection.

The outer frame is of interest because it contributes to the completion of the pattern and thus performs rather more than the usual structural function.

Ironwork in which this type of construction has been used does not entirely rely for its appeal on flow of line, and a silhouette effect. The surface of the metal acquires an attractive textural quality during forging, while chamfered edges, offset sections and rivet heads break up the play of light over the surface of the work and enhance the general effect.

The dimensions of this grille are 2′ 4½″ × 1′ 7½″.

DESIGN 3

Fig 49 The pair of tools illustrated are used for offsetting the 1″ × ¼″ bars at the intersecting points and junctions with the frame.

The guide pegs of the tool have been set at a distance which will just allow the bar to be moved freely through the tool without binding.

The bars are marked out in this instance at 4½″ centres and accurate positioning in the tool is ensured by aligning each centre as work proceeds with punch marks indicating the centre line of the bottom tool.

Fig 50 At a bright red heat the bar is positioned between the top and bottom tools and two or three light blows are sufficient to seat the heated metal in the bed of the bottom tool.

Excessive hammering must be avoided as this will lengthen the distance between centres, whereas these distances will remain unchanged if moderate controlled blows are used.

Fig 51 When marking out the section of bar from which the spurs are released it is essential to remember that all the natural chamfering left by the chisel cuts must be seen on one side of the work only, the face side.

It follows that the horizontal and vertical bars are not marked out on the same side, since the offsets are somewhat akin to halvings. Assuming that the horizontal bars are marked out with the impressed offset sections in a downwaıd position, the vertical bars must then be marked out on the opposite side, that is, with the impressed offsets in a raised position. (See this figure and *Design,* 4, Fig 67, page 44.)

The scheme of marking out is dictated by the fact that adequate sections of metal must be relieved from the bar in order to form the tapered quatrefoil sections but yet, at the same time, sufficient substance must be left in the parent bar to satisfy both structural and visual requirements.

The width of the bar is divided equally into three and the marking completed as shown in Fig 60 preparatory to hot chiselling. The essential points which determine the chiselling positions are marked with a centre punch. Two specially forged chisels are needed, one of each hand. (See drawing, Fig 62, page 39.)

Fig. 49

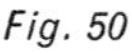

Fig. 50

Fig. 51

Fig. 52

Fig 52 The cut is made right through the hot metal from the face side only.

The cutting edge of the chisel is of the precise length and of the correct shape for the cut required. It must therefore be placed accurately and struck decisively. There is no room for fumbling.

A vessel containing water, in which the chisel may be cooled at intervals, must be within convenient reach.

A copper or soft iron plate must be used under the work to protect the edge of the tool from the anvil face.

A very slight lengthening of the parent bar takes place during this hot splitting process, which collectively affects the overall length of any given bar. The amount of lengthening can only be determined by making a short trial section.

DESIGN 3

Fig 53 At a red heat the work is engaged between anvil horns of the correct size (see drawing, Fig 63, page 39), and with the round-nosed pliers the spurs are moved away from the parent bar.

Fig 54 Using a stake (see drawing, Fig 64, page 40) specially designed to enable a number of operations to be done within a small compass, and a 'fuller' ended hammer (see drawing, Fig 65, page 41), a spur is drawn out to form part of the crescent shape which, with its pair, makes one quarter of a quatrefoil. When the smith's hand is lower than the part being forged, an old woollen glove is worn to ward off hot scale.

Since the section of the spur is not, at this juncture, square, and since the bevel left by the chisel must coincide with the face of the special stake during forging, distortion of the spur is inevitable.

Fig 55 If the work is viewed edgeways any distortion is readily seen and it must be dealt with in the course of forging.

Fig 56 The first stage in correcting distortion is carried out on the mandrel section of the special stake using the same hammer, the other end of which has been forged and ground to a flat square face.

The work is tilted at an angle and with well controlled blows the high edge of the spur, thrown into prominence in this position, is forged down from tip to root. By this means the spur is given the desired square section.

Fig 57 The spur—one only is dealt with at a time—is brought back into a plane true with the parent bar and its surface is levelled in one and the same operation. This work is done on the anvil-like surface of the multiple tool and, as in the three preceding operations, the length of the spur is increased.

When the forging of a pair of spurs has reached this point, a full quarter of a quatrefoil has been formed, and the tips should be long enough to connect with their neighbours when the grille is assembled.

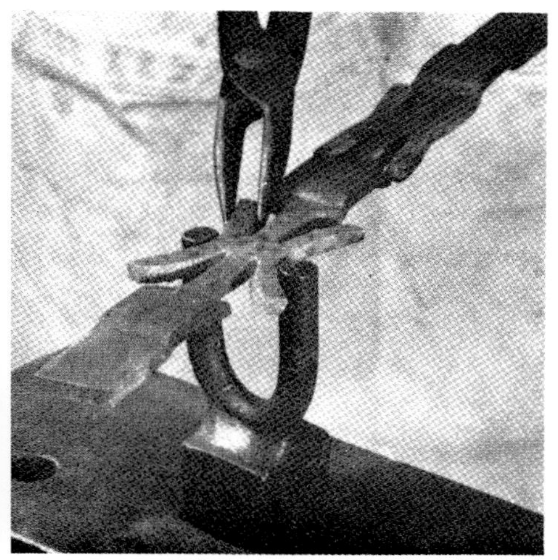

Fig. 53

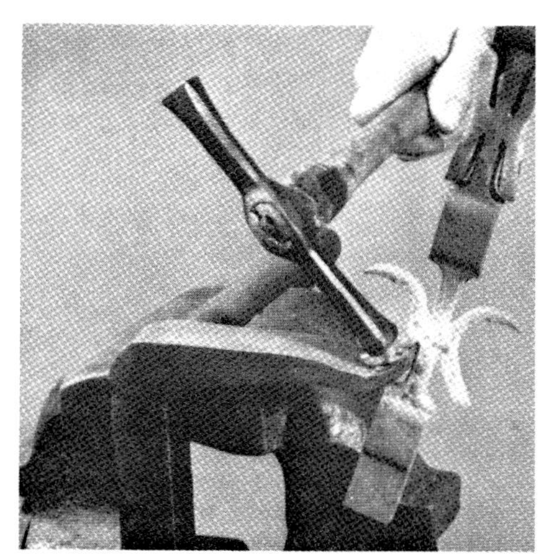

Fig. 54

Fig. 55

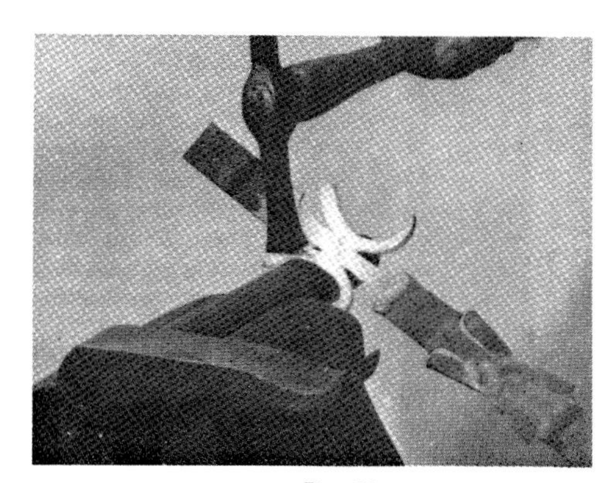

Fig. 56

Fig. 57

DESIGN 3

Fig 58 In the final setting operation another special tool is brought into play (drawing, Fig 66, page 42).

This tool comprises a crescent-shaped former and four guide pegs spaced to accept the cut-away stem bearing the quatrefoil sections.

With the work at a red heat, and using a pair of round-nosed pliers in each hand, the horns of the crescents are squeezed into shape against the former.

Fig 59 Providing that the offsets have been accurately positioned and that the correct reductions have been made between centres to allow for the slight lengthening which occurs during hot splitting, the assembly of the grille is straightforward.

As the frame bars bear crescent shapes on their inner edges only, they will not tend to lengthen as much as the internal bars, and therefore it is advisable to mark off the centres of their offset sections at the finishing intervals, i.e., the interval given by the working drawing.

Small, unpredictable variations can occur occasionally between centres in forgings of this kind, which must be adjusted prior to final assembly. Such adjustments may be effected by lengthening by the drawing process or shortening by upsetting, whichever is appropriate for the correction of the faulty sections.

The bars are straightened and any twists are removed, after which the rivet holes are drilled in the bars forming the outer frame.

The frame is temporarily bolted together at the four corners and tested for squareness.

The horizontal and vertical bars are now laid one at a time on the frame and the positions of their end holes are scribered off and drilled.

One set of bars, either vertical or horizontal, are now drilled at the points of intersection.

The grille is temporarily bolted together and the positions of the holes in the set of undrilled bars are scribered through, drilled and replaced in position.

At this stage, with frame and bars bolted together and again tested for squareness, any overlapping of the quatrefoil segments is corrected with a small scroll wrench.

Small, final adjustments to the tips of the segments are made after the temporary bolts have been withdrawn singly and replaced with rivets.

If it is found necessary to resort to the use of heat the oxy-acetylene torch is employed.

Fig. 58

Fig. 59

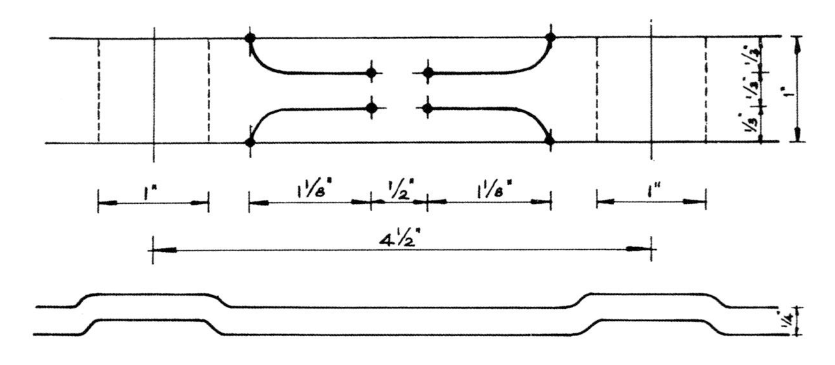

Fig. 60 Marking out diagram

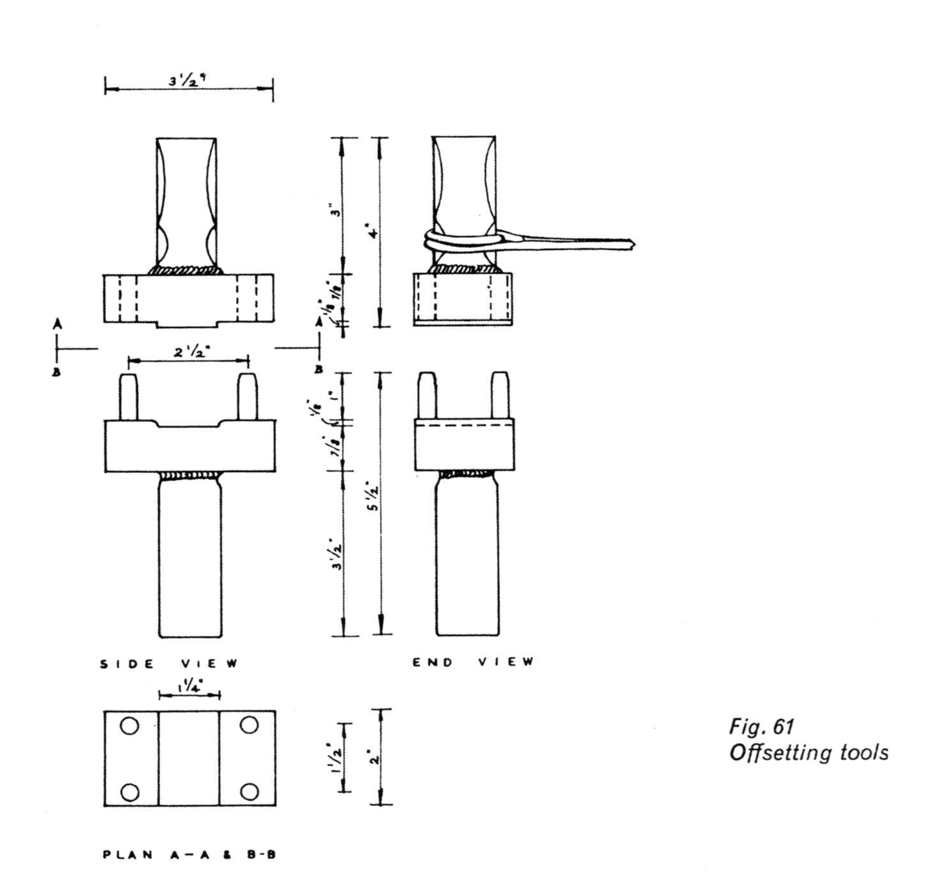

Fig. 61
Offsetting tools

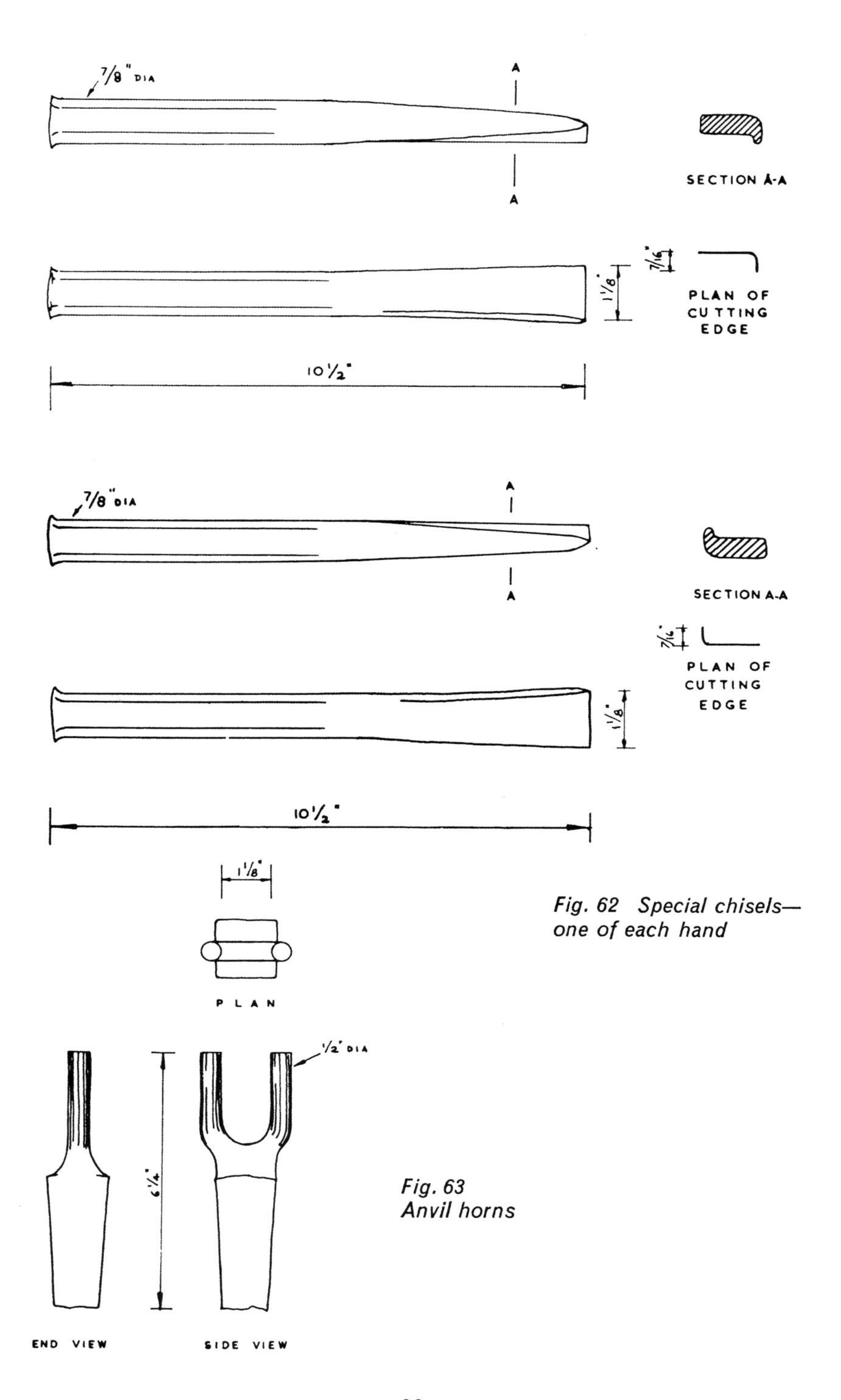

Fig. 62 Special chisels—one of each hand

Fig. 63 Anvil horns

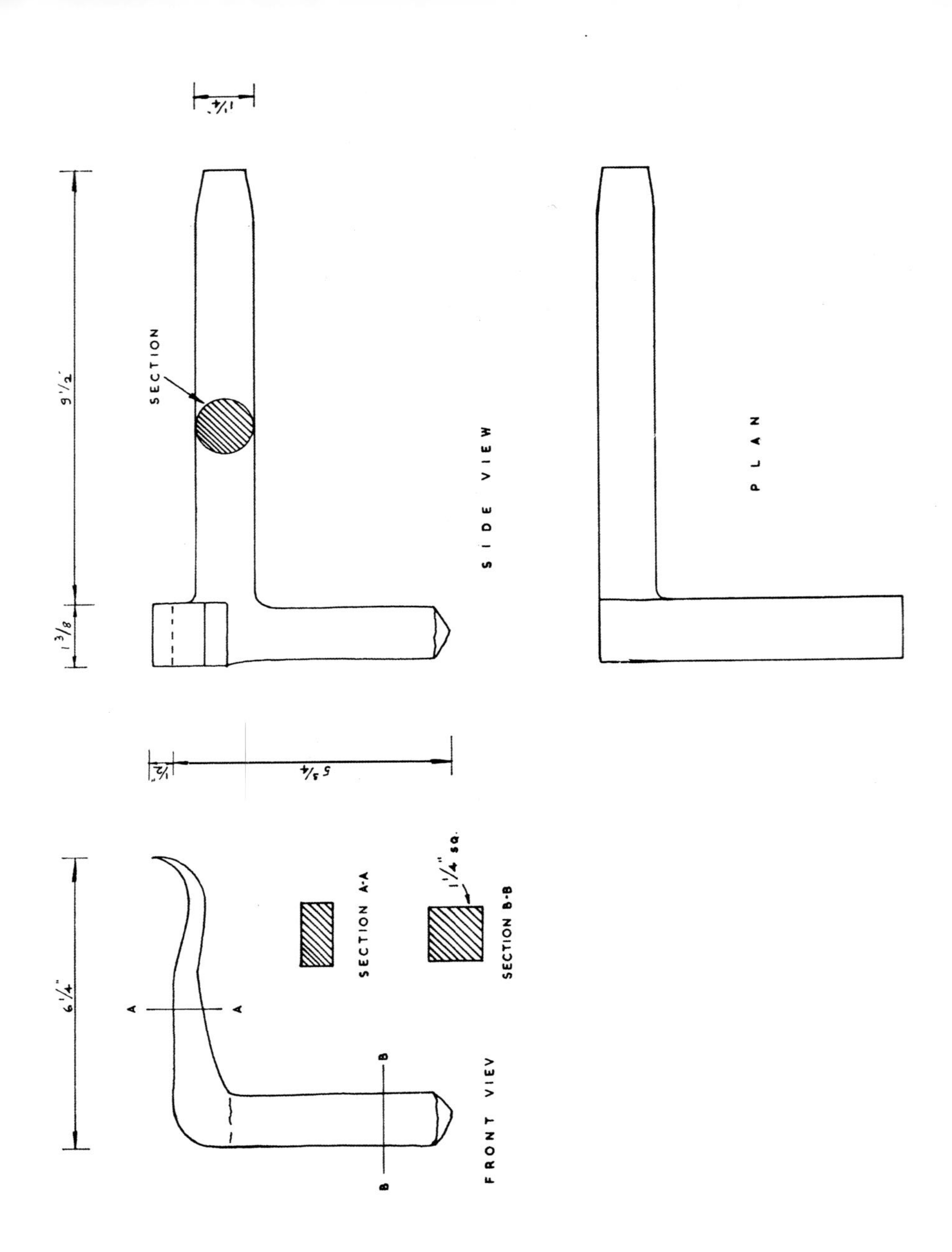

Fig. 64 Multi-purpose stake

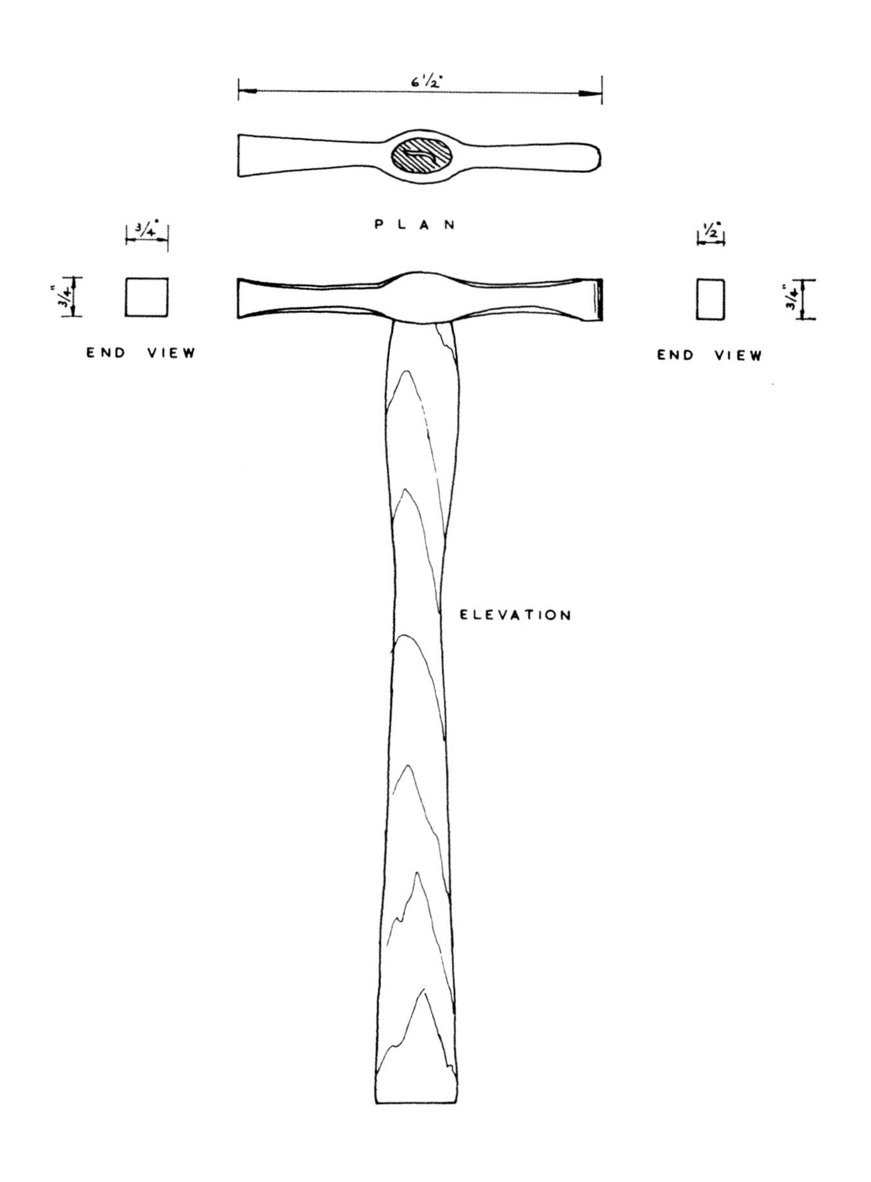

Fig. 65 Double-ended hammer

DESIGN 3

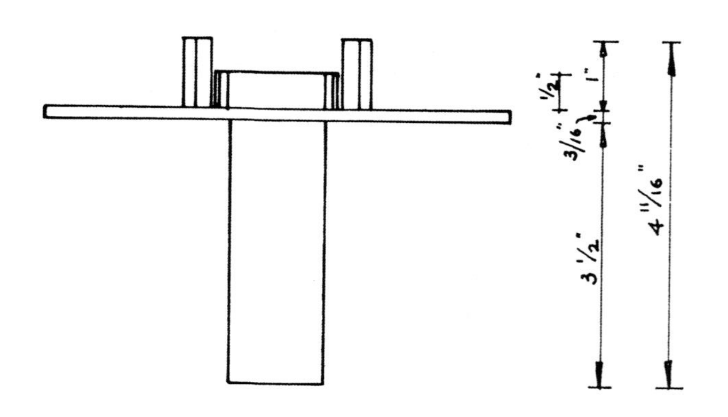

SIDE VIEW

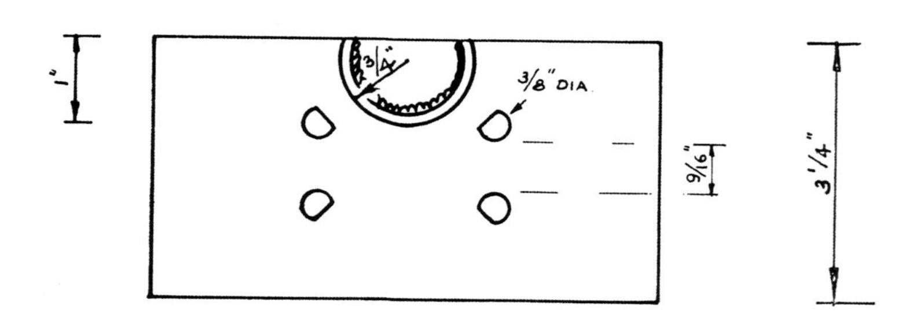

PLAN

Fig. 66 Jig for setting spurs

Design 4

A design similar in general character to No. 3, constructed in precisely the same manner: but differing from it in the way the decorative features are disposed.

In this case the spurs chiselled from the bars are not lengthened by forging, but are pulled away from the parent metal in gentle curves. The direction of the cuts have been reversed. They are started at a point midway between intersections and thus the curved spurs cluster around the intersection points when the grille is assembled, forming narrow petalled flowerlike motifs.

Since the main principles of construction are the same, it will be appreciated that the special tools used for the preceding grille (No. 3) will, to a great extent, be used in the present case.

Forging time is saved by the acceptance of the shape of spur left by the chisel cut, with little or no further dressing.

When designing work making use of this technique, it must be remembered that the shape of the remaining section of the parent metal is dependent on the original shapes of the portions cut from it. Thus the finished work will largely depend on a just balance between these two elements for its visual effect.

It is unnecessary either to describe or to illustrate the initial stage of construction, the offsetting process, as it is identical with that used in the case of Design No. 3, and it is similar in size, measuring 2′ 4½″ × 1′ 7½″.

DESIGN 4

Fig 67 The work is marked out for cutting (see drawing, Fig 73, page 47). The essential points determining the chisel positions are centre punched.

Left and right-handed chisels are not required as was the case in the previous example; one pattern of specially forged chisel with a curved cutting edge suffices. (See drawing, Fig 74, page 48.) The breadth of the soft iron or copper cutting plate should fit freely between the offset sections to support the bar when being cut, when the offsets are in a downward position, as well as to afford protection to the chisel edge.

Fig 68 Careful positioning of the chisel is particularly necessary so that the end of the cut which severs the edge of the bar just leaves the centre punch mark untouched. If this is not done, no accurate guide mark for making the corresponding cut on the same side of the bar is available. (For guidance on other points refer to design 3, Fig *52*.)

Fig 69 Using round-nosed pliers the spurs are curved away from the bar at a red heat, with the work between suitable anvil horns. Care must be taken when using the horns not to lever the bar laterally more than is necessary. Excessive leverage will distort the alignment of the centre portion of the bar.

The eye will soon become accustomed to judging the amount of curve required to achieve uniform setting, an advantage when the final shaping is done.

Fig 70 To give the spurs the correct smooth curve the multi-purpose stake held in the vice is used in a manner somewhat similar to that employed in fashioning the quatrefoil sections of design No. 3. This time, however, the fuller ended special hammer is not used to lengthen the spur but is lightly applied to induce the metal to assume the form required. Care is taken to avoid mutilating the natural bevel of the chisel cut. In this design the bevel is not forged away but is retained to add its quota to the overall decorative effect.

Fig. 67

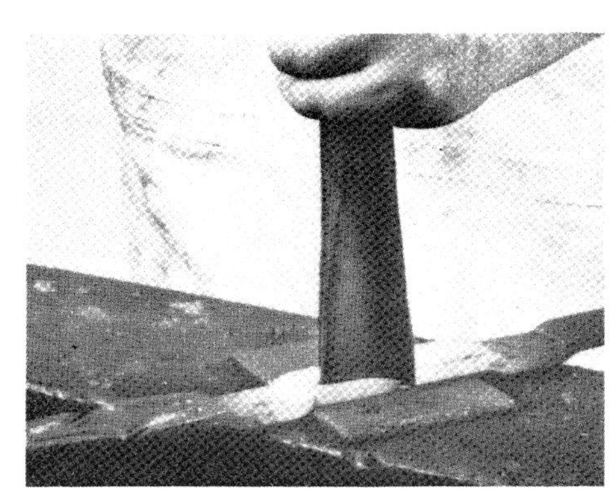

Fig. 68

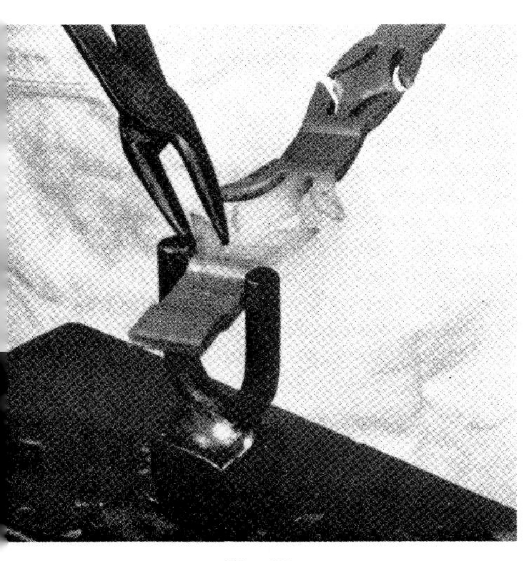

Fig. 69

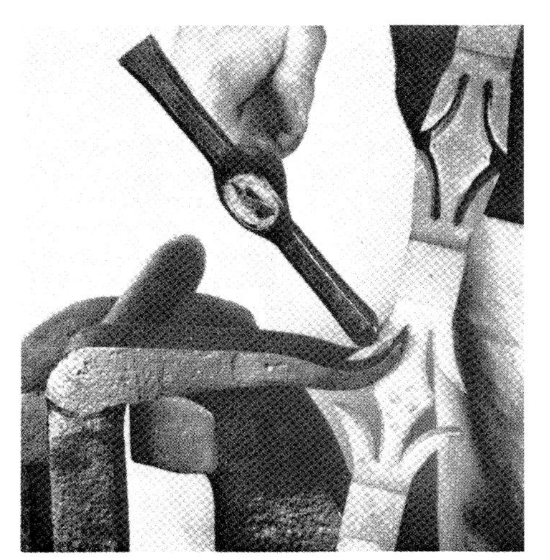

Fig. 70

DESIGN 4

Fig 71 Using the anvil section of the multiple tool the work is lightly hammered into a true plane.

It is necessary to check the work from time to time against a chalk tracing of the full size drawing on steel plate, in order to ensure that uniformity in setting has been achieved.

In view of the relative simplicity of the operations a special former is superfluous.

Fig 72 The assembling of this grille takes a similar course to the previous example, except that little or no time is absorbed in adjusting the spurs forming the decorative features.

It will be appreciated that the methods used for this and the previous example are flexible within certain limits and lend themselves to the creation of a range of designs. For instance, in the present case the spurs could have been shaped to bring the tips rather closer together, so forming flower forms of a different character. Furthermore, it is obviously possible to alternate flower forms of the two patterns, or even to embody both the quatrefoil motif of the previous example and one of the flower forms in the same grille.

There is also limited scope for the use of chisel cuts of differing profiles, but care must be taken to ensure that sufficient strength is left in the body of the bar.

Changes should not be made merely for the sake of novelty, as this approach might easily destroy character and produce nondescript results.

It is always very risky to accept ideas that have been worked out on the drawing-board only; a small section of the modified design should be made so that the decision whether or not to proceed may be based on observation of the effect in the solid.

Fig. 71

Fig. 72

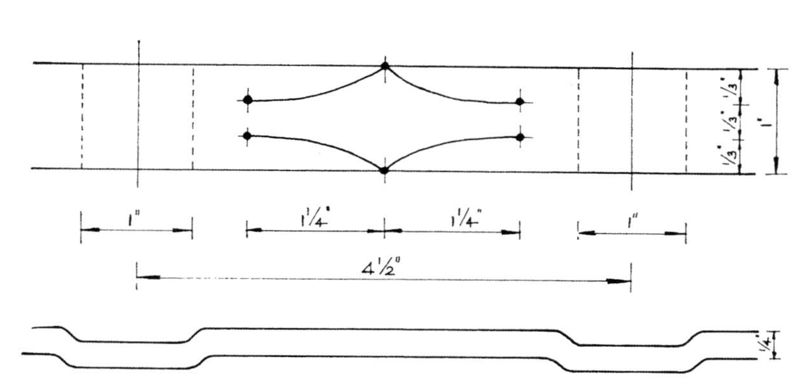

Fig. 73 Marking out diagram

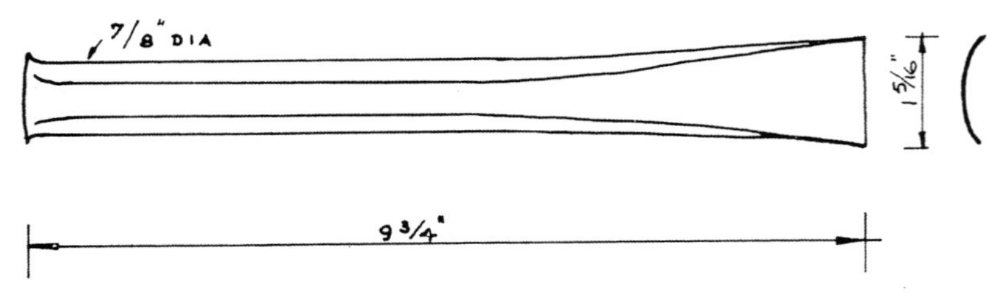

Fig. 74 Curved chisel

Design 5

In contrast to the other designs described in this book, the present example is, to a greater extent, built up from component parts applied to the bars, rather than fashioned from them.

The style of the decorative motifs also differs, since as the edge rather than the face of the metal is presented when the grille is viewed from the front. A delicate contrast is thus produced between curved forms and strong perpendiculars.

This design is particularly adaptable to settings of widely differing areas, a point which will become clear as the details of its construction are assimilated. It is not suggested that either the size of motifs, or the spaces between them, should be enlarged, which would produce an attenuated impression. A decrease or increase may be very desirable, but it is essential that a working drawing should first be made to ensure that the density of pattern is maintained and is appropriate to the setting.

The frame of the example illustrated is 3′ 4″ × 1′ $10\frac{1}{2}$″, and is made from $\frac{3}{4}$″ × $\frac{1}{2}$″ bar. The decorative motifs are of $\frac{3}{4}$″ × $\frac{3}{16}$″ and the vertical bars of $\frac{1}{2}$″ square material.

DESIGN 5

Fig 75 Work is started by taking a piece of $\frac{3}{4}'' \times \frac{3}{16}''$ bar of ample length to make one section of a decorative motif and bending the first corner to 110° using a hand lever bending machine.

Fig 76 The slight radius left on the corner by the bending machine is now converted into a sharp corner by forging. To do this a short heat is taken, which must be a balanced heat, that is equal in each leg of the angle. To ensure this result a brick is positioned near the fire on which one end of the piece is rested at the correct angle.

Fig 77 The corner is defined by working alternately on each of the outside faces. During this operation the work is kept clear of the anvil to avoid thinning.

Fig 78 When the hammer is applied in the second position, one leg of the angle is rested on the anvil against the step, where no slipping can take place.

Fig 79 A second corner is bent at 110°, at a distance from the first corner, found by measuring round the curve on the full size working drawing. This distance, once determined, remains constant throughout the work. To ensure complete control when defining the second corner, tongs of the correct pattern must be used.

Fig 80 The length of metal between the two corners is now formed into the correct curve. No heat is used; the operation is carried out solely with the aid of anvil horns and scroll wrench.

Fig. 75

Fig. 76

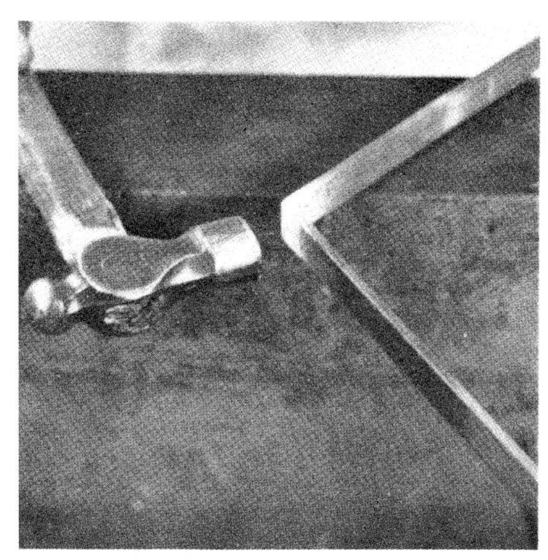

Fig. 77

Fig. 78

Fig. 79

Fig. 80

DESIGN 5

Fig 81 The curve, and set of the angles are checked against the drawing to ensure that both legs of the piece are in one true line.

Fig 82 The length of metal required for each leg, from one corner to the tip of the curled end, is now obtained from the full size drawing.

When the ends of the motif are fish-tailed by drawing down, lengthening of the metal will take place. This must be taken into account and an allowance made when the legs are cut to length prior to forging. The required allowance must be determined by trial for any given size of motif. After forging, the fish-tail is both trimmed to length and squared off in one and the same operation. The tip is curled over the front edge of the anvil by means of a stroking action with the hand hammer, in readiness for engaging on to a former which is specially made for this job. (Drawing, Fig 100, page 60.)

The engaged portion is held in place on the jig by gripping it with a pair of round-nosed pliers. The tip of the work must accurately coincide with the end of the jig before the operation of pulling the work round it is attempted.

Fig 83 The work is pulled into shape and if it has been correctly carried out the corner will coincide with the end of the straight part of the jig within close limits.

All measurements and details are checked during the process of making this first section to ensure that subsequent repetition work may go smoothly.

Fig 84 The decorative motifs are carried on collars made from round bar. In the present example they are of $1\frac{1}{8}''$ diameter. These may be made by cutting them from the bar with a power hacksaw and drilling them individually. Alternatively a lathe may be used, when the stock bar can be bored to a depth of several inches and the collars parted off.

A collar has been produced $1\frac{1}{8}''$ in diameter and $\frac{3}{4}''$ in thickness with a $\frac{7}{16}''$ hole drilled centrally.

The collar is heated and a $\frac{7}{16}''$ square drift is driven through the hole, over a bolster.

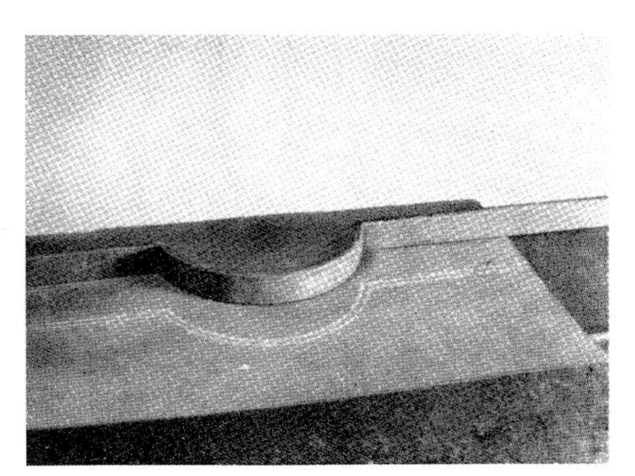

Fig. 81

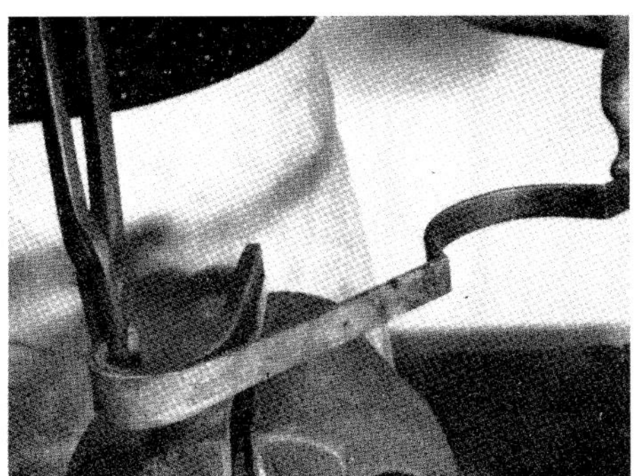

Fig. 82

Fig. 83

Fig. 84

DESIGN 5

Fig 85 With the $\frac{7}{16}''$ square drift in the hole, two facets, wide enough to accommodate $\frac{3}{4}''$ wide bar, are forged parallel to one another. The relationship between these facets and the axis of the drifted hole must be carefully observed from the illustration.

Fig 86 The drifted hole is inevitably distorted to some degree by the previous operation. This distortion is now rectified by forging the arcs of the collar employing a rolling action in a suitable bottom swage. Two purposes are served: the distortion is removed and the mechanical surface of the bar is given a hand-wrought texture.

Fig 87 The collar is now drifted with a $\frac{1}{2}''$ square clearance drift to accept a $\frac{1}{2}''$ square bar.

Fig 88 The faces of the collar are dressed in order to remove any irregularities left by the drifting and forging operations.

Fig 89 The vertical bars are now made. The centre bar is described here because it carries, in this case, two pairs of spurs not included in the other vertical bars. In a larger grille these features would occur on alternate verticals.

One end of a convenient handling length of $\frac{1}{2}''$ square bar is forged to a round section $1\frac{1}{2}''$ in length and $\frac{1}{2}''$ in diameter.

This section of the centre stem is set aside and work is continued by the making of a pair of spurs to be fire welded eventually to its rounded-up end.

Fig 90 A convenient handling length of $\frac{5}{8}''$ square bar is short square pointed leaving a small square flat area at the tip.

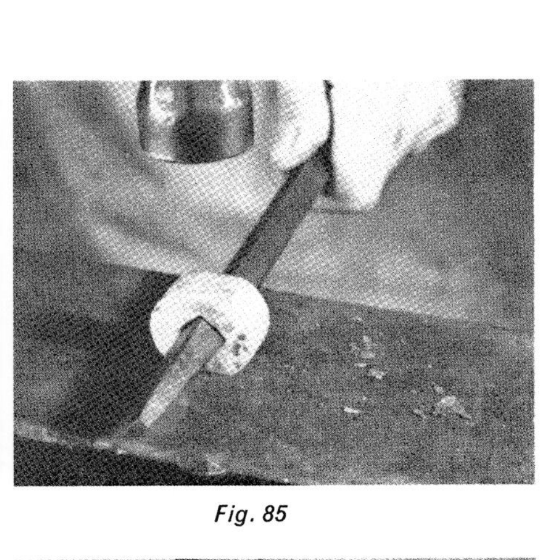

Fig. 85

Fig. 86

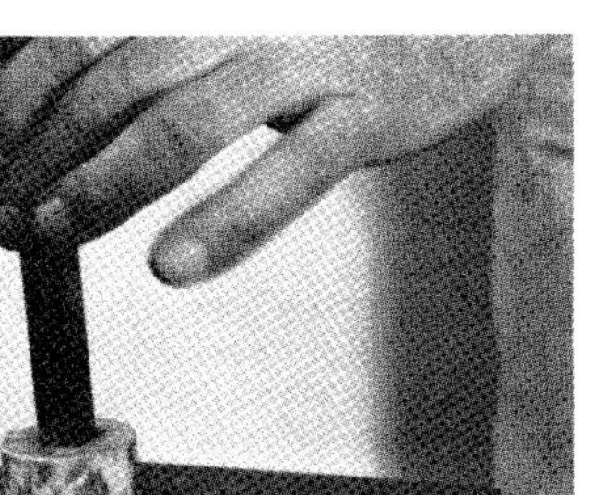

Fig. 87

Fig. 88

Fig. 89

Fig. 90

DESIGN 5

Fig 91 The pointed bar is now split with the hot-set across one diagonal to produce two tapering spurs of triangular section. The small flat area referred to above renders the positioning of the hot-set easier and ensures an accurate cut. The cut must travel down coinciding with the opposite corners of the bar. This point must be particularly carefully watched at the start, and to maintain regularity a rocking action in the same axis as the diagonal must be employed.

Fig 92 Splitting is continued sufficiently far to provide an adequate length, not only for the spurs, but also for an additional 1½″ which will eventually be necessary for fire welding the spurs to the bar. When sufficient length has been split, the two spurs are cut off from the bar.

Fig 93 Before the spurs can be welded on to the prepared end of the ½″ square bar, their ends for a distance of approximately 1½″ must be hollowed on the inside to embrace the round section.

To accomplish this a combination of fuller and 'V' block is used, the former to produce the required hollow and the latter to properly support the angled underside of the spur during the forging. The innermost end of the hollow formed by the fuller must not end abruptly, but must gradually diminish in depth to blend finally into the main form.

Fig 94 The first spur is tack welded into position. The spur and the bar are raised to a welding heat. When the bar and spur are withdrawn from the fire the former is placed in a bottom swage and the spur is positioned on the bar, care being taken that its rib falls in a true line with one of the edges and not one of the faces of the bar. The spur is firmly tacked to the bar with a few correctly tempered blows delivered by the smith's mate, who uses a curved faced hammer. He uses this particular hammer to prevent a sharp nick being made in the rib of the spur when delivering his first blows, which should fall near to the branching point and not on the end of the work. The aim is both to effect a good welded junction without reducing the section on the end of the bar, which must be left full for subsequent scarfing and to preserve, as far as possible, the rib of the spur.

Fig. 91

Fig. 92

Fig. 93

Fig. 94

DESIGN 5

Fig 95 One of the spurs has now been tack welded on to the bar whilst the other remains to be attached. This second spur is now welded on in the manner already described.

In practice as little time as possible would be permitted to elapse between operations. A prompt return of the bar to the fire for the welding of the second spur conserves heat and reduces scaling.

Fig 96 A full welding heat is taken and with a curved faced hammer the tacked spurs are finally welded to the parent bar.

Full control must be maintained at this point in order to blend in the edges of the spurs left proud and unwelded after the tacking stage, and to restore the edges which were inevitably bruised at that stage.

Fig 97 The work is dressed with 'V' shaped swages to correct any remaining irregularities.

Fig 98 An adequate length—in the present case about 2 feet—of ½″ square bar is fire welded on. It is particularly important that this weld should be perfectly sound because twists have yet to be made in the bar and this weld occurs within one of the twisted sections.

The weld has been left at an incomplete stage in order to illustrate the method employed.

Fig 99 The spur bearing bar is completed in the following manner.

The first twisted section, a few inches adjacent to the spur, is formed. As this is a tight or close twist, the work is done at a red heat.

A collar, correctly orientated and drilled and tapped to accept ¼″ Whitworth screws, is slipped on and the next twisted section is formed.

The second collar is slipped on in readiness for the following stage.

The second pair of spurs is forged and welded to a convenient length of bar by repeating the operation dealt with in Figs 89 to 97.

Fig. 95

Fig. 96

Fig. 97

Fig. 98

Fig. 99

DESIGN 5

This whole section is welded to length on to the main bar and the final twist is made.

After each twisting operation, work must be checked carefully to ensure that the alignment of collars, and spurs where they occur, is correct. Any inaccuracy in the setting of the collars will be magnified when the decorative motifs are attached to them and at that juncture adjustment may be difficult to effect.

In the case of uprights not carrying spurs, no welding is involved. Therefore, a start may conveniently be made at the centre twist.

A collar is slipped on after a twisted section has been formed and before the next is started. It must, in other words, be borne constantly in mind that these are not reduced twists.

The collars are secured in their correct position by means of screws which engage in notches made in the edges of the bars. The screws are of a length calculated to tightly engage these notches and at the same time firmly secure the motifs.

The frame of the illustrated example is made with the broad faces of the bars outwards. Therefore, the ends of the inner bars are flattened to provide riveting surfaces.

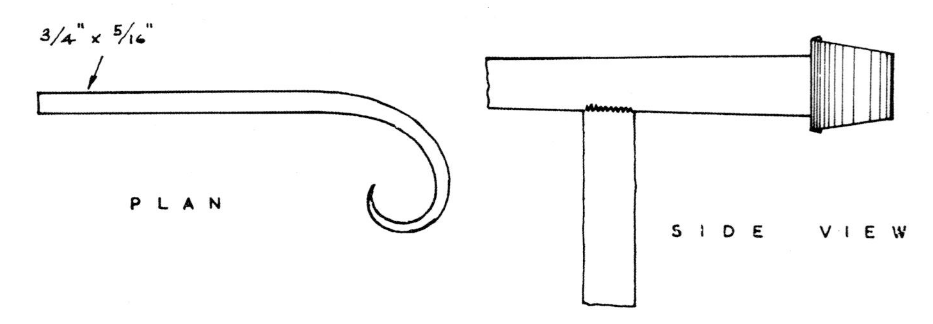

Fig. 100 Jig for forming curled ends

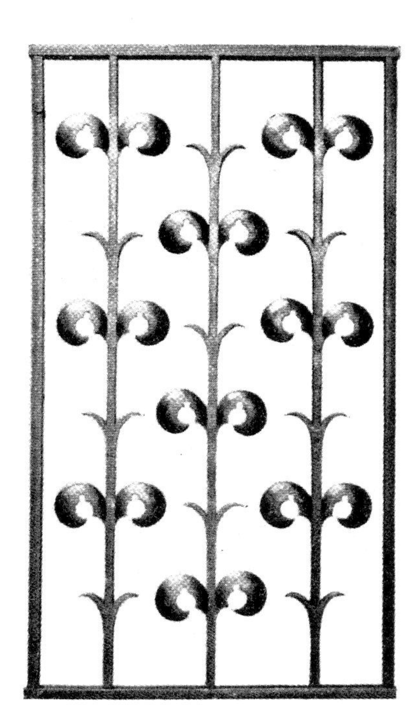

Design 6

The decorative effect of this design depends to a large degree on the effective disposition of two contrasting motifs, one forged in the traditional manner and the other formed by combining handwork with the use of machinery.

Opportunity to exploit the latter practice does not frequently present itself, and even when it does arise the advisability of following such a course must be most carefully considered.

On the other hand, however, when the assistance of machinery is appropriate, time may be saved and economic considerations can, in the modern world, be of importance and may, in fact, be the deciding factor.

The description of the technical operations which follows includes no dimensions, as proportions and motifs may be modified according to the perception of the designer, or by the dictates of a specific commission.

DESIGN 6

Fig 101 To fashion the spur motif a piece of metal long enough for a pair of spurs is drawn down at each end. It must be remembered that this piece must include enough metal for fire welding below the branching point.

A central impression is made with a fuller to facilitate bending. The piece is bent and folded on to the first section of the central stem which is a convenient handling length.

The branching points are now marked with a centre punch for guidance when welding.

The spurs are now removed from the centre stem which is heated to a bright red heat, quickly withdrawn from the fire and reinserted in to the folded spurs.

A light welding heat is taken to secure the spurs to the stem preparatory to taking a full welding heat. The spurs must not be allowed to slip during this operation.

Fig 102 Where an arc welding plant is available the spurs may be made separately and tack welded into position ready for the fire welding operation. Preheating of the centre stem is impossible. It is therefore necessary to ensure that sufficient heat reaches and penetrates into the stem without the spurs becoming overheated. Thus this method involves rather more soaking in the fire, but there is, on the other hand, no danger of the spurs slipping out of place.

Fig 103 A full welding heat is taken and a curved faced hammer is used to weld the spurs to the stem, back to the centre punch marks. This work is done on the anvil bick, which not only automatically leaves the end of the metal thick in readiness for scarf welding on the next section of the stem, but also prevents any reducing of the spurs which would result if the anvil face were used. Thorough welding is essential because any weakness would allow so short a length to pull away from the stem when the spurs are set to pattern.

Fig 104 The next section of the stem is scarfed and fire welded on. As this section will be cut to length after welding, at the point where

Fig. 101

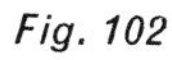

Fig. 102

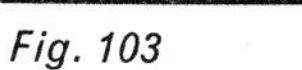

Fig. 103

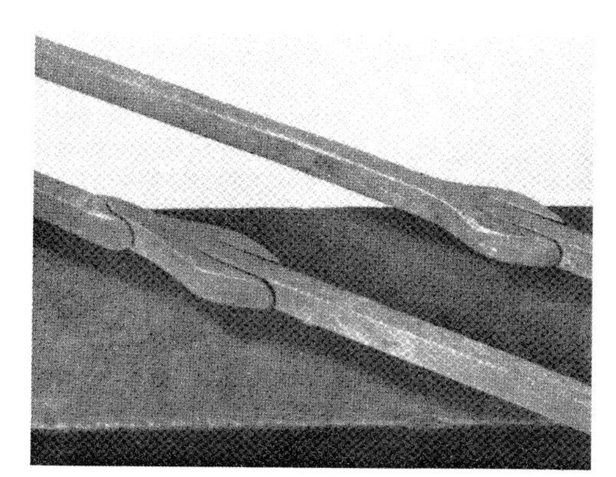

Fig. 104

the next pair of spurs is to be attached, an ample length of metal must be used. It will be noted that in the nearest example displayed on the anvil face the weld has been left unfinished in order that the juncture of the scarfed ends may be observed.

The position of the next pair of spurs is marked to conform with the working drawing and any surplus stem is cut off. These spurs are now welded on and the sequence of operations is repeated as many times as necessary.

DESIGN 6

Fig 105 Spurs of so short a length are difficult to set unless a stake of special design is employed. The shape of the required stake is illustrated (Fig 64, page 40). This is gripped in the vice and the spurs are curved to shape with a hammer designed to reduce this and similar operations to single-handed work, obviating the use of a fuller and the consequent employment of a mate.

Fig 106 The blanks for the leaf features are next cut from sheet steel $\frac{1}{8}''$ in thickness. The design employed permits part of their outlines to be formed either by the punching or the boring of a number of holes of equal diameter. Punching with a fly press is illustrated; alternatively blanks may be stack piled and drilled.

After the series of holes is either punched or bored the remainder of the outline is trimmed away. In the illustrated example it will be noted that additional notches were made in order to break the regularity of the circular holes.

Fig 107 At a red heat the edges are hammer chamfered using a curved faced hammer on one side of the work only.

Fig 108 The leaf motifs are now dished by beating them, at a dull red heat, on a concave surface of lead in a steel socket of suitable diameter. During this operation the lead is never allowed to accumulate sufficient heat to melt it. The inner and outer edges of the cylindrical socket should have been filed to produce a rim of rounded section.

During the beating process, changes in outline develop as the contours of the motif are formed. It therefore follows that the final form to be achieved is dependent on the initial outline of the blank in which suitable allowances must have been made.

The development shape may be determined by first modelling the motif in plasticine. A thin sheet of this material is rolled out, formed and subsequently carefully flattened to give the guiding outline. A trial made in sheet metal will either confirm the result or will reveal points where modification is needed. Lead sheet could be used for the same purpose but plasticine is recoverable.

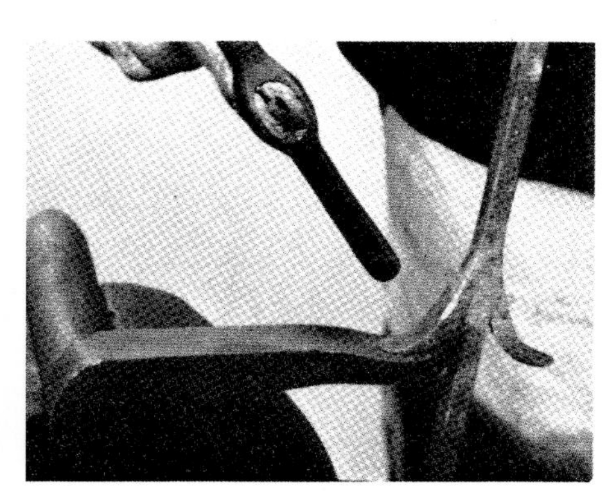

Fig. 105

Fig. 106

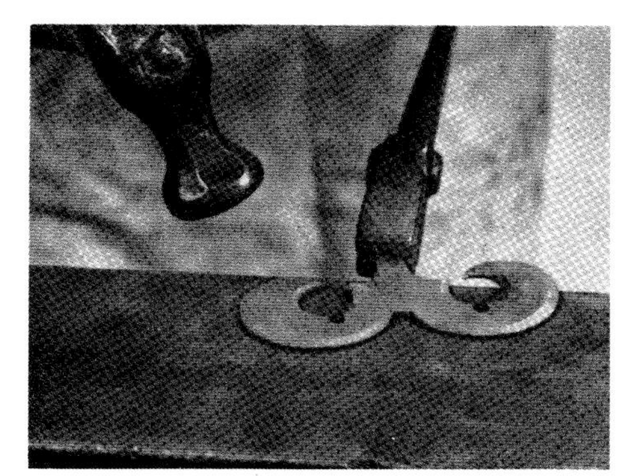

Fig. 107

Fig. 108

DESIGN 6

Fig 109 The flattened surfaces bisecting the motifs are trued with a hand punch (drawing, Fig 113, on page 68) of correct width. This operation prepares the correct seating areas matching the width of the bars to which they are attached.

Fig 110 The intervals in the bars are now accurately marked out and recessed to the correct depth to accommodate the flattened areas of the motifs.

Fig 111 Flush riveting the motifs to the bar is the next operation and if it is considered desirable the junctures between the sheet metal and the parent bar may be lightly hammered-up to conceal them.

Alternatively the skilful use of bronze welding could be instrumental in the saving of time without impairing the visual effect of work.

Fig 112 Riveted tenon construction has been used to assemble the grille illustrated, but alternative methods may be adopted such as the form used in several other grilles described in this book.

In some instances work must present a similar appearance from both sides. The design dealt with in this chapter has, as it stands, a face side and reverse. Conversion to double sided work may be effected by facing the other side with additional leaf motifs, back to back wise, carried on rivets common to the pairs.

Fig. 109

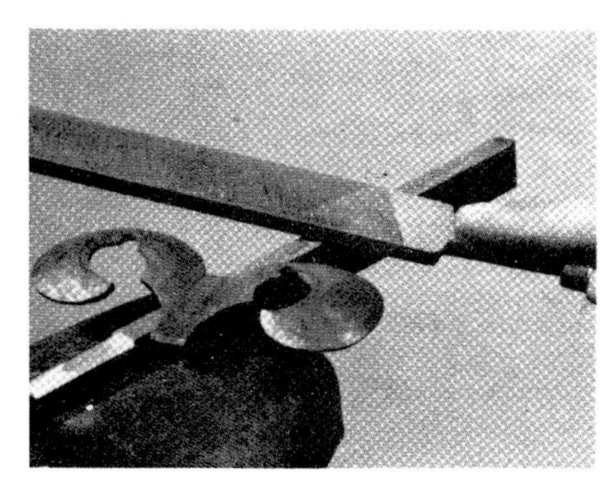
Fig. 110

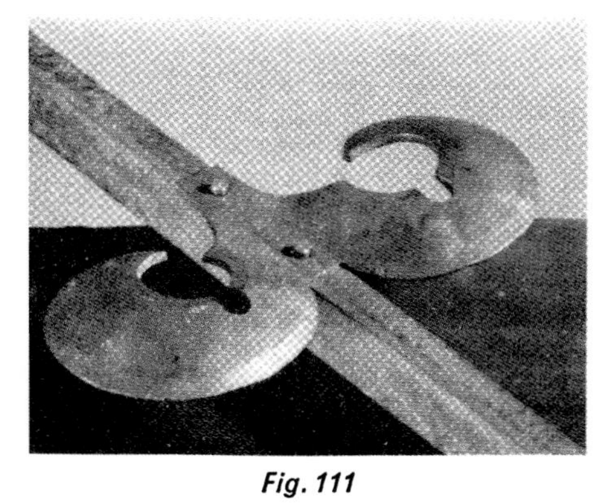
Fig. 111

Fig. 112

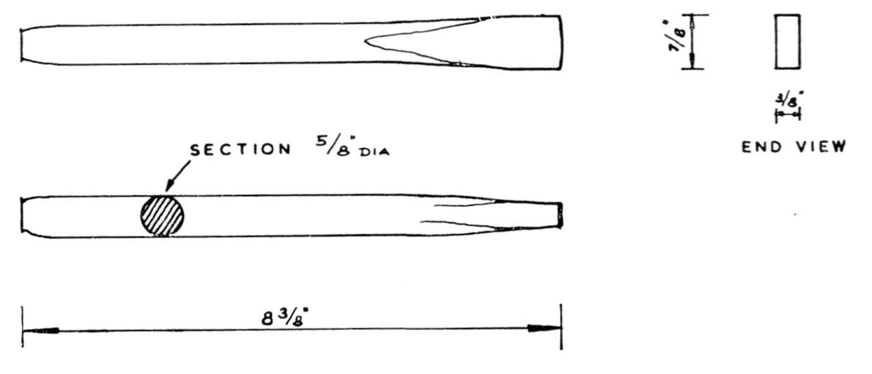

Fig. 113 Hand punch of correct width

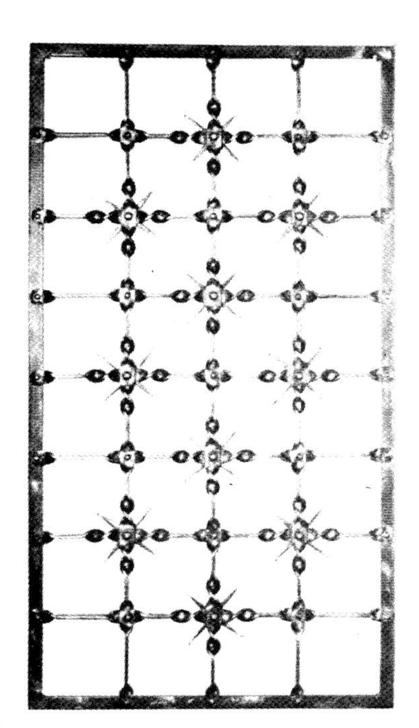

Design 7

The decorative features in this example are simple in character and are formed by the intersection of cold twisted bars.

The design may be carried out in bar sizes ranging from $\frac{1}{2}'' \times \frac{3}{16}''$ to $1'' \times \frac{1}{4}''$. If sizes in excess of $1'' \times \frac{1}{4}''$ are used, then hot twisting must be resorted to with its attendant technical complications.

The size of bar chosen must be closely related to the division of the given area into squares, if the flower-like features are to tell with full decorative effect. The use of too narrow a bar will, for instance, result in the features becoming insignificant. Therefore, in describing the processes involved, measurements relating to the bar size used in the illustrated example, $\frac{3}{4}'' \times \frac{3}{16}''$, are detailed and given in drawing, Fig 125, on page 75.

It is essential that a trial of any given bar be carried out in the first instance for the craftsman's or designer's guidance when setting out work employing this technique. An adequate working drawing cannot be made if this preliminary step is omitted.

The frame of the illustrated grille is of $\frac{3}{4}'' \times \frac{3}{8}''$ bar, $2' 11\frac{1}{2}'' \times 1' 6\frac{1}{2}''$ overall, giving an opening of $2' 10'' \times 1' 5''$. The bars, $\frac{3}{4}'' \times \frac{3}{16}''$, intersect at $4\frac{1}{4}''$ centres, producing a satisfactorily rich effect, neither meagre nor too overpowering in feeling.

No further reference will be made to the frame as it is assumed that its construction presents no problem to the skilled worker.

DESIGN 7

For the same reason the riveted assembly is not mentioned.
The formation of the twisted bars, however, requires description.

Fig 114 All horizontal and vertical bars are cut to adequate lengths, care being taken to include a surplus to accommodate the wrench adequately during the formation of the final twist.
In this instance the horizontal bars are 1′ 10″ long, giving a 4″ surplus.
To make the top horizontal bar, with its large centrally placed star-decorated motif, flanked by the two smaller motifs, the bar is marked in the middle with a centre punch and gripped in a vice dog with this point coinciding with a mark made centrally on one jaw (drawing, Fig 126, page 76).
The width of the vice dog jaws forms a gauge for one of the essential measurements in this pattern; the length of the flats at crossing points.
A wooden gauge is made to predetermined measurements, i.e., taken from the full size working drawing. Using the widest leg of the gauge, 1¾″, the wrench (see drawing, Fig 127, page 77) is positioned and tightened on to the bar.

Fig 115 A three-quarter twist, 270°, is made in a clockwise direction.
The wrench is removed and the work released from the vice dog, turned end for end and the operation is repeated.

Fig 116 One half of the larger star-bearing motif has now been formed. The ⅞″ flat surface left between the twists will become a riveting point when crossed at right angles by a similarly twisted bar.

Fig 117 **It might be assumed that twisting operations materially shorten the length of the bar. Shortening does take place but in the quarter twist, 90°, which follows, it is negligible. The distance to the next riveted intersection point may, therefore, be marked with the dividers set at precisely 4¼″.**

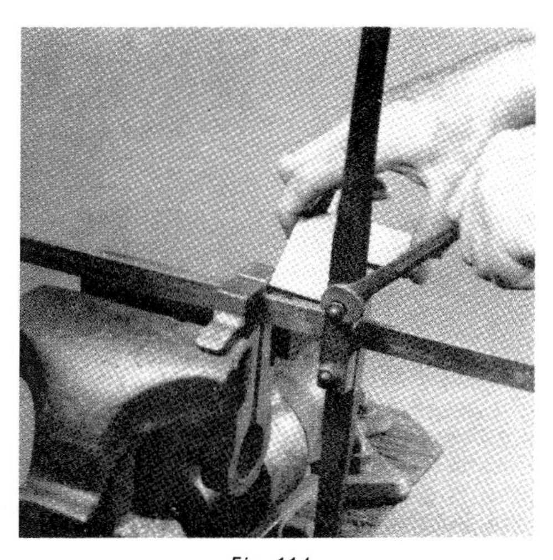

Fig. 114

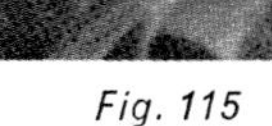

Fig. 115

Fig. 116

Fig. 117

This is a convenient juncture at which to emphasise the necessity of marking the bar as each successive twist is made. By so doing any slight shortening resulting from twisting need not be taken into account.

DESIGN 7

Fig 118 The bar is gripped with the mark in the centre of the vice dog and using the small leg, $\frac{5}{8}''$, of the gauge the wrench is tightened in position as before. Care must be taken when aligning the marks.

Fig 119 A quarter (90°) twist is made, but this time in an anti-clockwise direction. The wrench is removed by taking out one bolt, and the work is released from the vice dog, turned end for end and the operation is repeated.

One half of the smaller motif has now been formed. As before, the $\frac{7}{8}''$ flat surface becomes a riveting point.

Fig 120 It will be noticed that the operation carried out in Figs 117-119 have been repeated on the other side of the larger twisted motif, leaving only the final twists at each end of the bar to be made.

These twists do not form part of intersections, but make junction with the frame. Therefore the measurement required to position the wrench is taken from the centre of the bar and is half the inside measurement of the frame, in this case $8\frac{1}{2}''$. This point is marked, and coincides with the inner edge of the frame. The wrench is, therefore, tightened with its innermost edge to this mark.

The work is now entered into the jaws of the vice dog and the bar is moved towards the dog. The vice is tightened when the gauge, $\frac{5}{8}''$, fits between the dog and the wrench.

One quarter (90°) twist is made in an anticlockwise direction.

Fig 121 The ends of each bar are heated, rounded and hammer chamfered; they are cut off short to allow for the lengthening which will take place.

Fig 122 The finished bar ready for drilling.

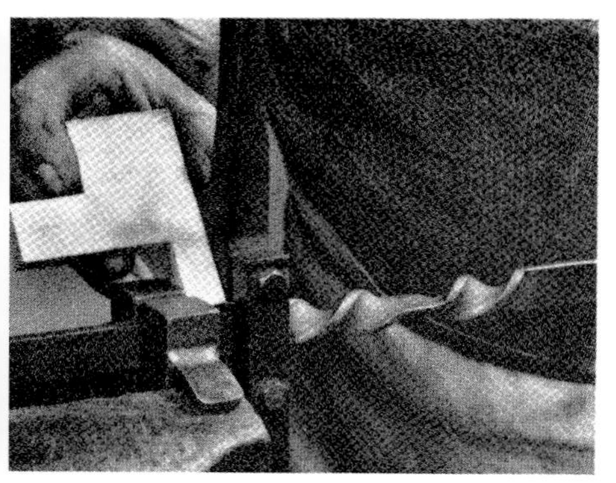
Fig. 118

Fig. 119

Fig. 120

Fig. 121

Fig. 122

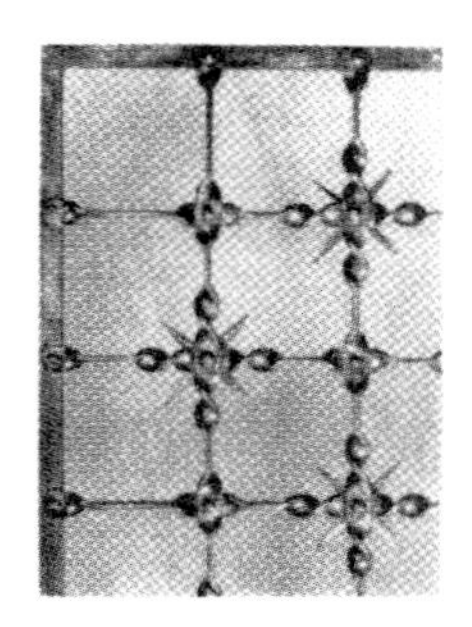

DESIGN 7

Fig 123

The four-way stars are made from $\frac{3}{8}'' \times \frac{1}{8}''$ bar to a size proportionate to the overall design. A handling length of bar is taken and drawn down to a point, care being taken to observe accurately the amount of draw which occurs. The point is hammer chamfered along its edges on both face and undersides. The piece is now cut off at a point taking into account the length of draw already determined, and similarly pointed at the other end.

Two more points are made which are cut off to a length which will produce a balanced feature when assembled by means of oxy-acetylene welding.

Fig 124

The remainder of the bars, both horizontal and vertical, are made by repeating the operations described, the sequence varying as dictated by the design. Thus alternate bars will start in the centre with the smaller motif flanked by the star-bearing motif.

When the grille is riveted together the star features are not placed between bars but at the back of the intersection.

Fig. 123

Fig. 124

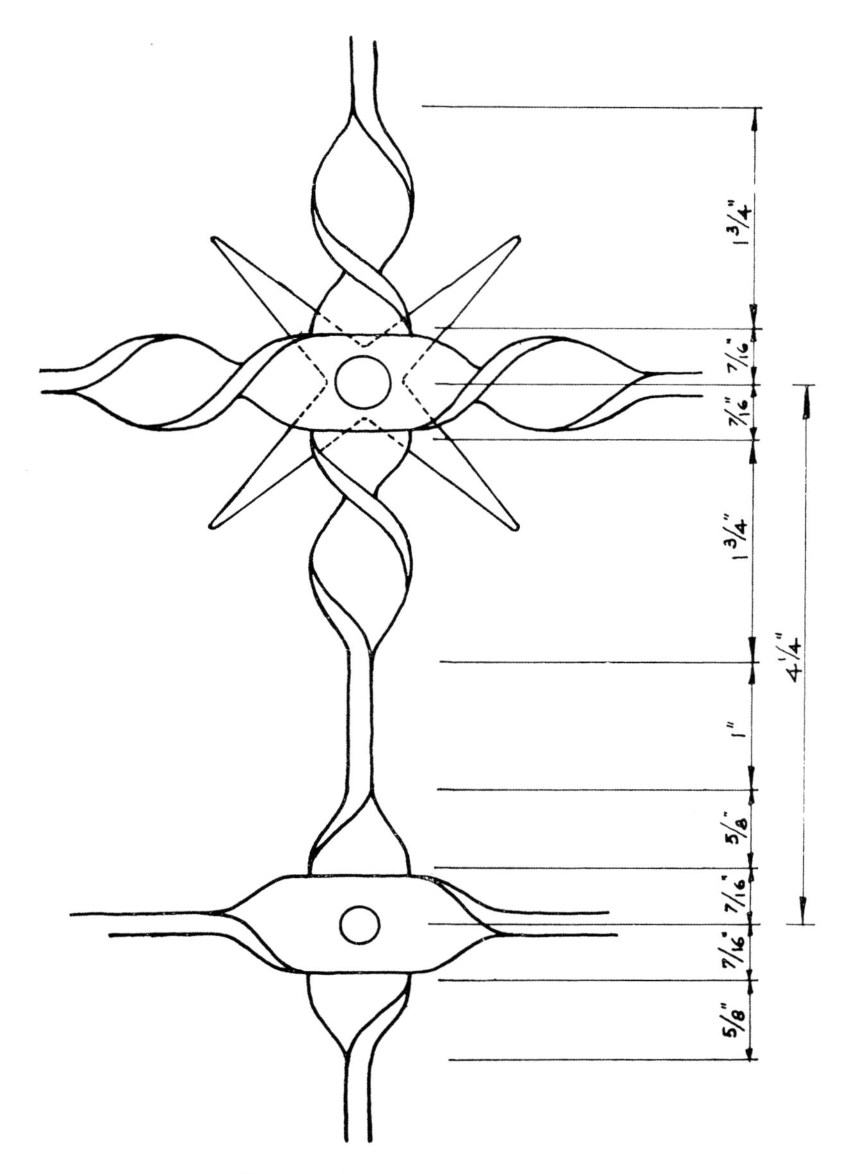

Fig. 125 Working measurements

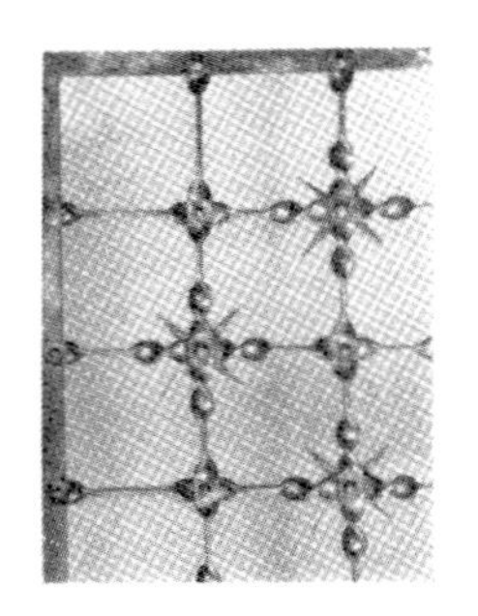

DESIGN 7

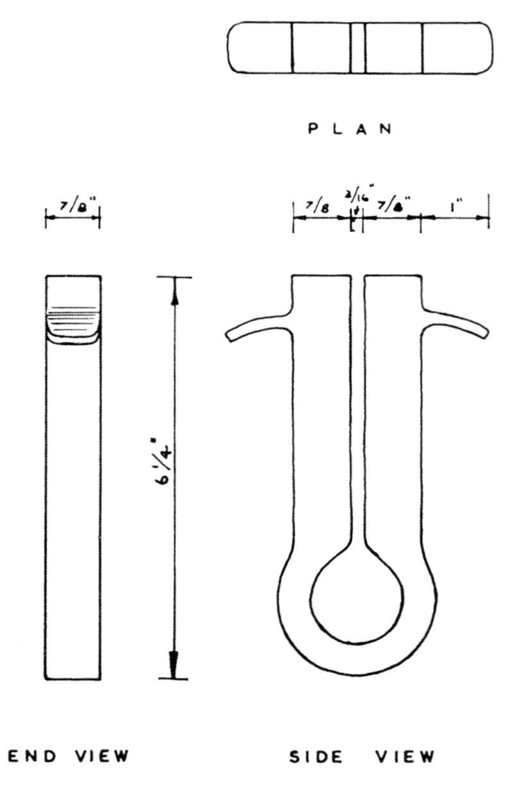

Fig. 126 Vice dog

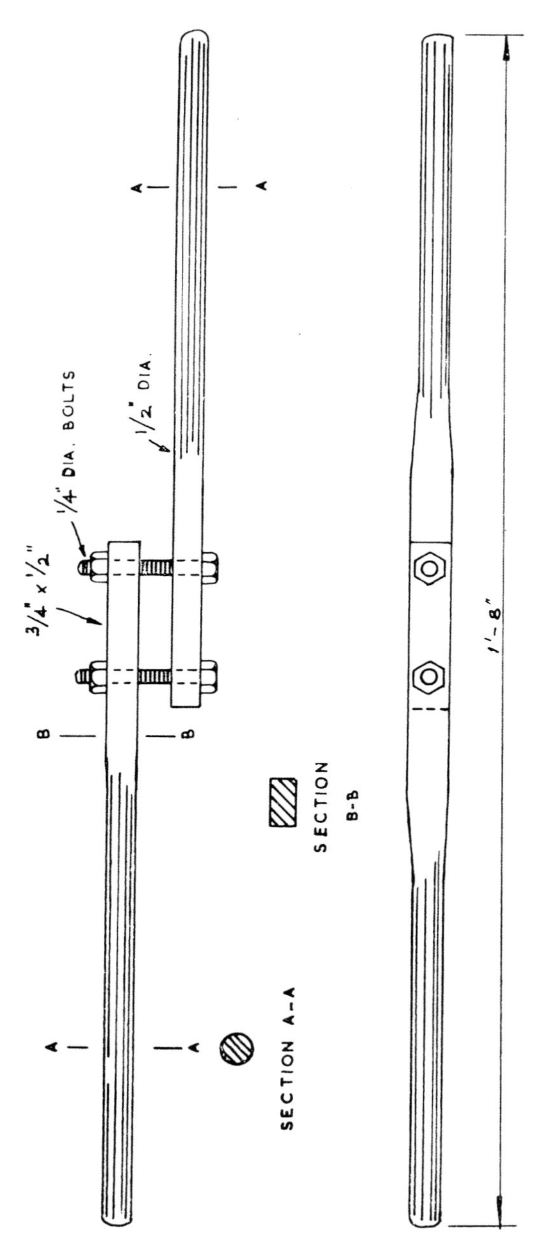

Fig. 127 Twisting wrench

INDEX

Other Great Metalworking Books:

BACKYARD FOUNDRY FOR HOME MACHINISTS
978-1-56523-865-7
$12.99 (US) $16.99 (CAN)

THE HOME BLACKSMITH
978-1-62008-213-3
$19.99 (US) $27.50 (CAN)
£14.99 (UK)

BLACKSMITH'S CRAFT
978-1-4971-0046-6
$14.99 (US) $18.99 (CAN)
£9.99 (UK)

METAL LATHE FOR HOME MACHINISTS
978-1-56523-693-6
$12.95 (US) $14.99 (CAN)

WROUGHT IRONWORK
978-1-4971-0064-0
$9.99 (US) $13.99 (CAN)
£6.99 (UK)

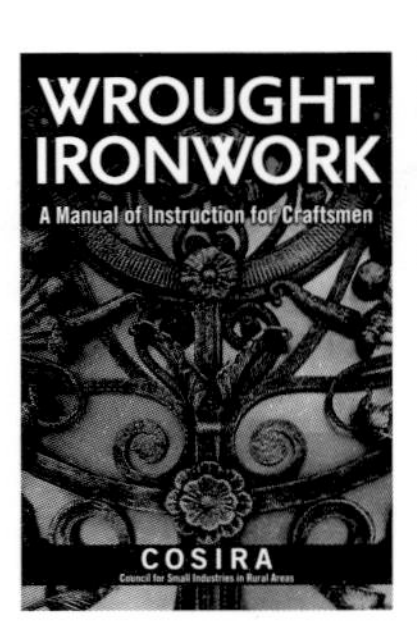

METALWORKER'S DATA BOOK FOR HOME MACHINISTS
978-1-56523-913-5
$14.99 (US) $17.99 (CAN)

FARM AND WORKSHOP WELDING
978-1-56523-741-4
$24.99 (US) $29.99 (CAN)

THE METALWORKER'S WORKSHOP FOR HOME MACHINISTS
978-1-565623-697-4
$14.99 (US) $17.99 (CAN)